TURBIDITY IN THE AQUATIC ENVIRONMENT

Publication Number 1057

AMERICAN LECTURE SERIES®

A Publication In

The BANNERSTONE DIVISION *of*
AMERICAN LECTURES IN
ENVIRONMENTAL STUDIES

Consulting Editor

CHARLES G. WILBER, Ph.D.
Department of Zoology-Entomology
Colorado State University
Fort Collins, Colorado

TURBIDITY IN THE AQUATIC ENVIRONMENT

An Environmental Factor
in Fresh and Oceanic Waters

By

CHARLES G. WILBER, Ph.D.

*Department of Zoology-Entomology
Colorado State University
Fort Collins, Colorado*

CHARLES C THOMAS • PUBLISHER
Springfield • Illinois • U.S.A.

Published and Distributed Throughout the World by
CHARLES C THOMAS • PUBLISHER
2600 South First Street
Springfield, Illinois, 62717, U.S.A.

This book is protected by copyright. No part of it may be reproduced in any manner without written permission from the publisher.

© *1983 by* CHARLES C THOMAS • PUBLISHER
ISBN 0-398-04726-X
Library of Congress Catalog Card Number: 82-6000

With THOMAS BOOKS *careful attention is given to all details of manufacturing and design. It is the Publisher's desire to present books that are satisfactory as to their physical qualities and artistic possibilities and appropriate for their particular use.* THOMAS BOOKS *will be true to those laws of quality that assure a good name and good will.*

Printed in the United States of America

I-R-1

Library of Congress Cataloging in Publication Data

Wilber, Charles Grady, 1916-
   Turbidity in the aquatic environment.

   (American lecture series; publication no. 1057)
   "A publication in the Bannerstone Division of American lectures in environmental studies."
   Bibliography: p.
   Includes index.
   1. Turbidity. 2. Hydrology. I. Title.
II. Series.
GB665.W556 1982   574.5'263   82-6000
ISBN 0-398-04726-X   AACR2

PREFACE

THE turbidity of natural waters was for a long time a characteristic that was taken as observed without attributing too much significance to it. Turbid water was viewed as dirty water or as favorable or unfavorable for fishing depending on one's personal viewpoint. More recently turbidity has been viewed as a diagnostic property of water that may reveal important information about water quality, changes, origins, and ecological impact. With the advent of remote sensing in limnology and oceanography the importance of the optical properties of water has been emphasized.

Absorption and scattering can be measured with great precision by remote sensors; suspended particulate matter in natural waters is responsible for those two characteristics. Water color which is measurable at a distance by modern techniques of color photography tends toward the blue in clear water but shifts toward the red in waters that are more turbid. These color characteristics can reveal important information about a given body of water.

The usual books on oceanography and limnology do not devote much space to turbidity. The present short volume is an attempt to put together in one place a summary of information on turbidity as a property of water that has environmental significance. The book should be useful and of interest to a varied readership: managers of environmental programs, students involved in aquatic studies, individuals who wish to carry out a quick familiarization program on turbidity as an environmental variable in the aquatic medium.

In even such a modest endeavor as this one an author is

indebted to many persons and institutions from which he has been given aid and support. Over the years, I have had consistent and sympathetic support from all sections of the university family at Colorado State University. The excellent climate provided at the University for intellectual freedom makes it an attractive milieu for creative scholars. The office staff in the Department of Zoology and Entomology, Colorado State University, have uniformly given me clerical, stenographic, and secretarial support in an outstanding and cheerful manner. The library personnel at The Marine Biological Laboratory, Woods Hole, Massachusetts have been over the years of crucial help; I appreciate especially how rapidly and effectively they have been able to complete demanding literature searches for me. The understanding and forbearance of my wife and family in the face of my attempts to be a scholar are deeply appreciated.

The personnel of the Charles C Thomas, Publisher, have in their usual easy manner been supportive and helpful.

Charles G. Wilber, Ph.D.
Fort Collins, Colorado

ACKNOWLEDGMENTS

THE extensive literature generated over the past decade on turbidity and in the ocean and fresh waters makes it not practical to an exhaustive, all-encompassing review with a true bibliography. This present work is based on a selection of published material that reflects the overall state of aquatic turbidity research at the end of 1980, when the literature review was terminated. Titles of articles were retrieved by querying two separate data bases through the computer services of the Marine Biological Laboratory, Woods Hole, Massachusetts. I gratefully acknowledge the invaluable help of the Library staff at M.B.L.

CONTENTS

Page

Preface v

Chapter

   1. Introduction and Physical Considerations. 3
   2. Some Ecological Effects 18
   3. Biological Effects 25
   4. Methodology 57
   5. Dynamics of Suspended Particles 67
   6. Estuaries 73
   7. Geographic Studies 89
   8. Waste Disposal, Fresh Waters, Abatement,
      Literature Reviews. 109

Bibliography. 117

Index 131

TURBIDITY IN THE AQUATIC ENVIRONMENT

Chapter 1

INTRODUCTION AND PHYSICAL CONSIDERATIONS

Definitions

ACCORDING to the *Glossary of Oceanographic Terms* turbidity is "Reduced water clarity resulting from the presence of suspended matter. Water is considered turbid when its load of suspended matter is visibly conspicuous, but all waters contain some suspended matter and therefore are turbid" (Baker et al., 1966).

Another term that arises in connection with a discussion of turbidity is "suspended sediment." This term has been defined by Faye et al. (1980) as sediment carried in suspension by the turbulent components of the fluid or by Brownian movement.

Sorensen et al. (1977) maintain the following with respect to the freshwater environments: "Natural surface or groundwater is never found as pure H_2O. Separation of the impurities of natural water into particulate and dissolved fractions is, in practice, made on the basis of working definitions such as those found in Standard Methods . . . . Suspended solids are the residue in a well-mixed sample of water that will not pass a standard (glass fiber) filter. The residue trapped on the filter is dried (103-105C) and reported in units of weight per volume (mg/1). Suspended solids usually impart an optical property to water called turbidity. Particulate matter causes light to be scattered and absorbed rather than transmitted in straight lines. This property (turbidity) can be measured by standardized methods but it cannot be related to weight concentrations of suspended solids because of the effects of size, shape, and refractive index of the particles. However, turbidity measurements do give an indication of the relative abundance of suspended material in a water sample."

Some attempt is made to classify the particle size of different types of sediments whether suspended or deposited. The classification follows:

Sediment Classification

Class	Size, mm
Boulders	256
Cobbles	64-256
Gravel	2.0-64
Sand	0.062-2.0
Silt	0.004-0.062
Clay	0.00024-0.004

Turbidity Current

This phenomenon consists of a turbid relatively dense current composed of water and sediment that flows downslope through less dense seawater; the sediment eventually settles out forming a turbidite (Ross, 1977). The matter of turbidity currents is still somewhat controversial in various details.

Useful General References

An understanding of oceanic turbidity, its nature, origin, and interpretation will be aided by reference to several excellent books that are available on optical studies of the ocean: Jerlov, 1968; Williams, 1970; Jerlov and Nielsen, 1974; Tyler and Smith, 1970; Frummeter and Mertens, 1975; Preisendorfer, 1976.

Timofeeva in Chapter 9 of Jerlov and Nielsen (1974, pp. 178-219) presents a well-documented discussion of the optics of turbid waters based primarily on laboratory experiments. The results of laboratory studies apparently agree so well with observations made in the ocean under "natural" conditions that Timofeeva recommends the "wider use of the method of laboratory investigations in optics of turbid waters and in other fields of research associated with the study of light in scattering media" (Jerlov and Nielsen, 1974, p. 217).

Some Basic Properties of Water

Measurements of refractive index of water from the mercury line at 4046.6Å to the helium line at 7065.2Å indicate that for all wavelengths the refractive index increases with decreasing temperature; at longer wavelengths this increase is maximal between 0-1°C (Eisenberg and Kauzmann, 1969). Moreover for a specific temperature the refractive index of water is a little less for the longer wavelengths of light.

If a layer of pure water has a beam of single-colored light (visible range) sent through it, most of the light is (a) transmitted through the water, or (b) reflected from the surfaces encountered. A small percentage of the light becomes scattered from the original direction of the beam.

The Rayleigh ratio, R, is the measure of this scattering:

$$R_\mu(\lambda) = \frac{I\lambda\, d^2}{I_o}$$

where $I\lambda$ is the intensity of the light scattered by a specific volume of the water; I_o, the intensity of the incident light that goes through the sample; λ is the angle of the incident light beam; d, distance from sample and locus at which I_o is measured. Subscript μ specifies that the incoming light is not polarized.

The scattering of light by liquids generally stems from two separate sources:

(a) "The incident beam induces oscillating dipole movements in the molecules, and the oscillating dipoles act as sources of secondary light waves" (Eisenberg and Kauzmann, 1969). The frequency of the scattered waves is the same as that of the incident light.

(b) "The anistropy of polarizability of molecules causes some additional scattering" (Eisenberg and Kauzmann, 1969). The depolarization ratio is the ratio of the intensity of the horizontal component to that of the vertical component of polarized light of a beam scattered in at 90° direction. From this depolarization ratio the additional scattering referred to above can be ascertained experimentally.

The Rayleigh ratio for light scattering measured experimentally falls between 2.30 and 2.9 10^{-6} cm^{-1}, at d = 436m; at d = 546m the range is 0.84 to 1.08 10^{-6} cm^{-1}. Experimental values for the depolarization ratio at d = 436m fall between 0.083 and 0.1, at d = 546m, 0.076 and 0.116 (Eisenberg and Kauzmann, 1969).

For further discussions on the structure of water as it may affect the optical properties see Eisenberg and Kauzmann, 1969; Frank, 1970; Van Hylekama, 1979; Cerofolini and Cerofolini, 1980; and Beall, 1981.

Brightness Coefficient

The coefficient of spectral brightness in the ocean is a value that describes the energy distribution in the spectral radiation coming from the sea. It thus characterizes the intrinsic color of the ocean (Kozlyaninov, 1979). This coefficient can be measured in by so-called noncontact procedures. Kozlyaninov (1979) maintains that this spectral brightness coefficient depends "only on the primary hydro-optical characteristics and the conditions of radiation passing through the sea surface which in many cases are rather stable." There is apparently a significant correlation of the brightness coefficient with the coefficient of diffuse reflection from the ocean and the absorption coefficient, and the vertical extinction coefficient. Essentially it can be demonstrated mathematically that the brightness coefficient of the sea depends exclusively on the ratio between backscattering and absorption of light.

Scattering

The relationships between the chemical, physical and optical properties of water in an organically rich, highly turbid area has been described by Thompson and coworkers (1979). They found that "scattering by suspended particulate material was found to be the primary optical mechanism controlling downwelling irradiance at all three wavelengths"; i.e. 445, 542, 630nm respectively.

Optical Properties of Seawater

The optical properties of seawater have been used to estimate the amount of chlorophyll and of suspended matter in the upper layers of the ocean (Bekasova, Kopelevich and Sud'bin, 1979). It is pointed out that "the spectral values of the brightness of upward radiation are used to determine the optical properties of seawater and the chlorophyll and suspended matter concentrations in the surface layer of the ocean." It was found that the most desirable circumstances for making observations of this sort include a cloudless sky with the sun at an elevation of 3 to 60 degrees and the sea condition no greater than 4. If the elevation of the sun is greater than 60 degrees, bright spots in the field of the measuring instrument make the observations fluctuate in a significant manner. Direct measurements of the concentration of chlorophyll and of suspended matter were made during one of the routine cruises of the R/B Vityaz. Backscattering was measured using a "spectrohydronephelometer." The results indicated that the brightness spectra may be used for estimating chlorophyll concentration in the surface layers of the ocean. It is important, however, to realize that the data obtained in this particular study apply only to waters in which the concentration of chlorophyll is moderate to low. The authors refuse to speculate on what sort of variations will be obtained in highly productive waters and in waters in which there is significant vertical variation of chlorophyll distribution. The authors found some disagreement between the measured and theoretical absolute values for the brightness coefficient. At the present time they are unable to explain this situation.

Attenuation

Information about dissolved and suspended matter in natural waters can be revealed by spectral characteristics of upwelling radiation, which has gone through downward in a column of water into the deep sea and then is backscattered. Appropriate sensors can pick up the upwelling signal. In order to reap the potential harvest of information available from this phenomenon, it is necessary to understand the relationship between the optical qualities of the ocean and the biological factors that modify these qualities.

Smith and Baker (1978) have reported a method that relates "the spectral diffuse attenuation coefficient for irradiance to the chlorophyll-like pigment concentration in a way that is both physically and biologically meaningful."

The mean concentration (to depth of 1 attenuation length) of chlorophyll a and pheopigments (milligrams of pigments/m^3), C_k, is related quantitatively to the optical factor, K_T, that relates spectral irradiance just beneath the ocean surface to the downwelling spectral irradiance at a (wavelength). C_k is related, as well, to a description for the color of the ocean. These relationships form the basis for remote-sensing of concentrations of chlorophyll-like pigments in oceanic waters.

K_T, the diffuse attenuation coefficient for irradiance, is a measure of the "bio-optical" condition of oceanic waters (Smith and Baker, 1978). Analyses of K_T and pigment concentration in seawater show that specific attenuation attributable to chlorophyll-like pigments is 0.016 ± 0.003 [m^{-1} (mg pigment m^{-3})$^{-1}$]. K_T is readily sensed by appropriate spacecraft instrumentation; it is strongly correlated with C_k, the concentration of chlorophyll-like pigment in the water. From K_T and C_k the fraction of radiant energy attenuated by phytoplankton can be obtained. These relations seem to open the door to effective monitoring of important aspects of marine ecosystems from satellites in a rapid, repeatable manner.

The following equations are of interest in connection with the two papers by Smith and Baker (1978, 1978b):

a. $K_T^{-1} = 8.78 - 7.51 \log C_k$
b. $C_{1T} = -0.20 + 0.788 \log C_{1k}$, $r = 0.95$
c. $K_T^{-1} = 9.46 - 7.90 \log C_k$, $r = 0.81$
d. $\log C_{2k} = -0.708 + 0.729 \log C_{1k}$, $r = 0.91$
e. $\log P_T = 1.254 + 0.728 \log C_k$

K_T^{-1} meters, Attenuation length
C_k, mg chl a/m^3, chlorophyll a concentration
C_{1T} = mg chl a/m^3 in euphotic zone
$C_1 K$ = mg chl a/m^3 in depth zone of 1 attenuation length
$C_2 K$ = mg pheopigments/m^3
P_T = primary productivity, mg C/12 hr./m^2

Energy absorption by ocean water varies according to depth and wavelength (Table 1-I), (Figures 1-1, 1-2, 1-3).

Particle Size

The study of particle size of materials suspended in sea water and the size distribution of particulate material in the ocean may be made more effectively if one is familiar with the recent publication by Bagnold and Barndorff-Nielsen (1980). These authors point out particle size distribution is frequently illustrated best by a log-histogram. If the log-histogram is nearly parabolic, the Gaussian or so-called normal distribution applies. There are a significant number of log-histograms of particle-size data that are hyperbolic rather than parabolic. The authors present a clear and useful description of hyperbolic distributions; they show how widely applicable such distributions are.

Works of sedimentologists are also useful in planning observations and in evaluating data related to size of particulate matter suspended in the ocean, e.g. Socci and Tanner, 1980.

Optical and Biochemical Comparisons

Carpenter and Carpenter (1979) have compared optical and biochemical methods for classifying ocean waters. They used a multivariate statistical analysis to evaluate Jerlov's optical classification of ocean waters and to compare that classification with one that was developed from other oceanographic measurements. According to these investigators: "It is shown that the water masses sampled can be allocated to their optical water types using criteria related to levels of primary production." (Carpenter and Carpenter, 1979).

Light Measurements

A number of investigators have concerned themselves with the derivation of various coefficients used in optical studies for light in the seawater. Kozlyaninov (1979) studied the luminescence coefficient in seawater. The effect of different kinds of scattering was described by Televin et al., 1979. The problem of attenuation in the Pacific Ocean was also discussed by Televin and coworkers (1978). Light-scattering and extinction of light in very turbid coastal waters was discussed by Thompson et al., 1979.

10 *Turbidity in the Aquatic Environment*

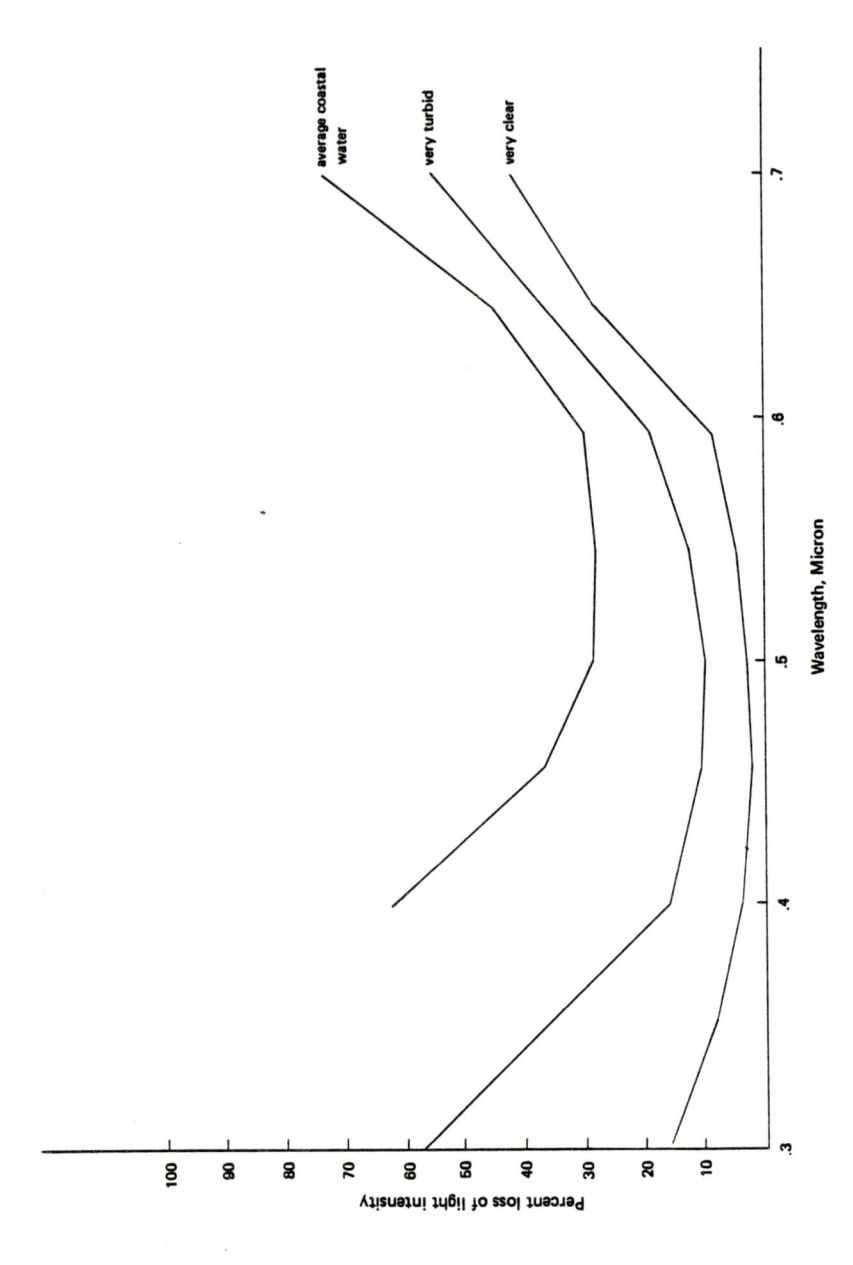

Figure 1-1. Percent loss of intensity of light passing through 1 meter of sea water of different clarity.

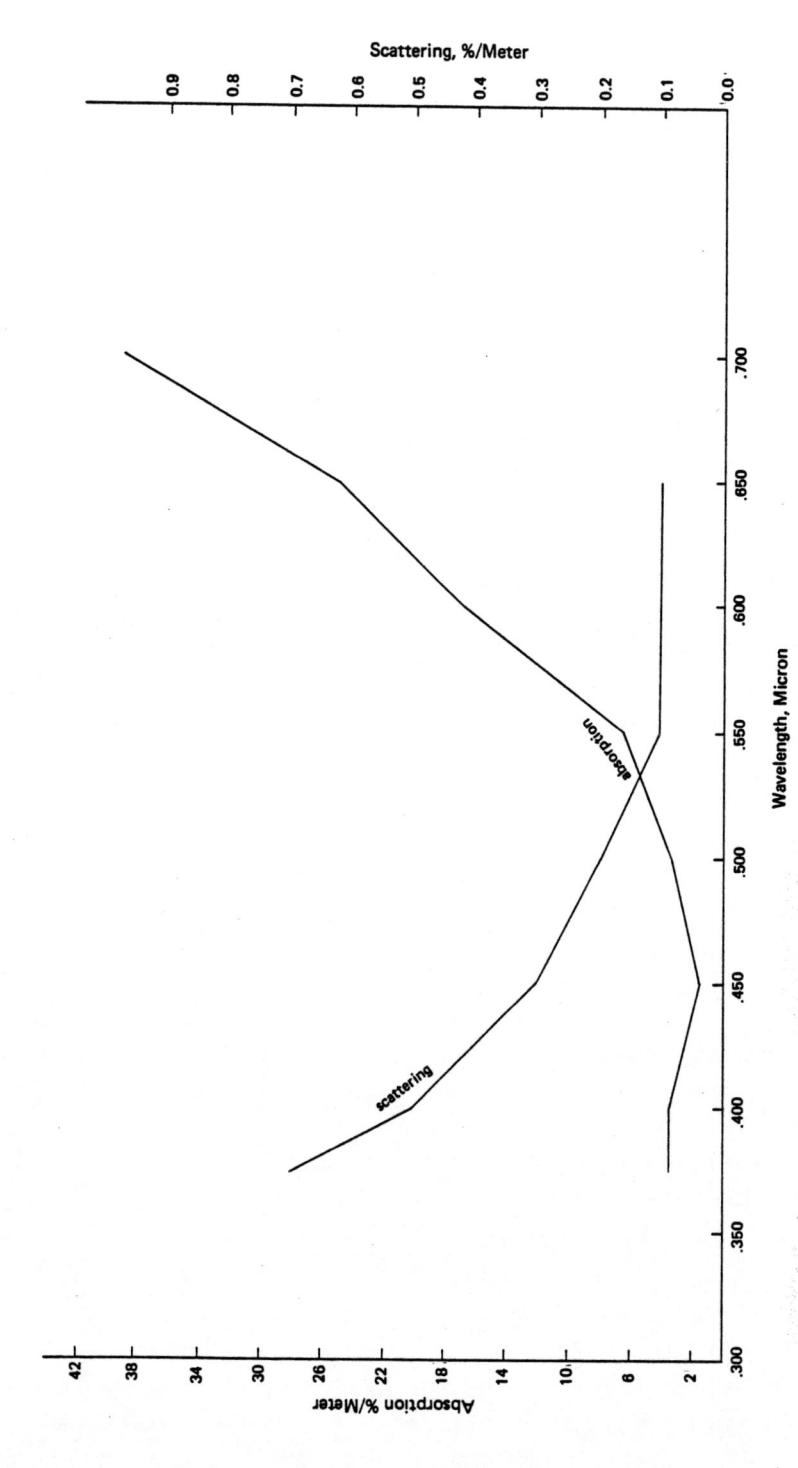

Figure 1-2. Attenuation of light in pure seawater.

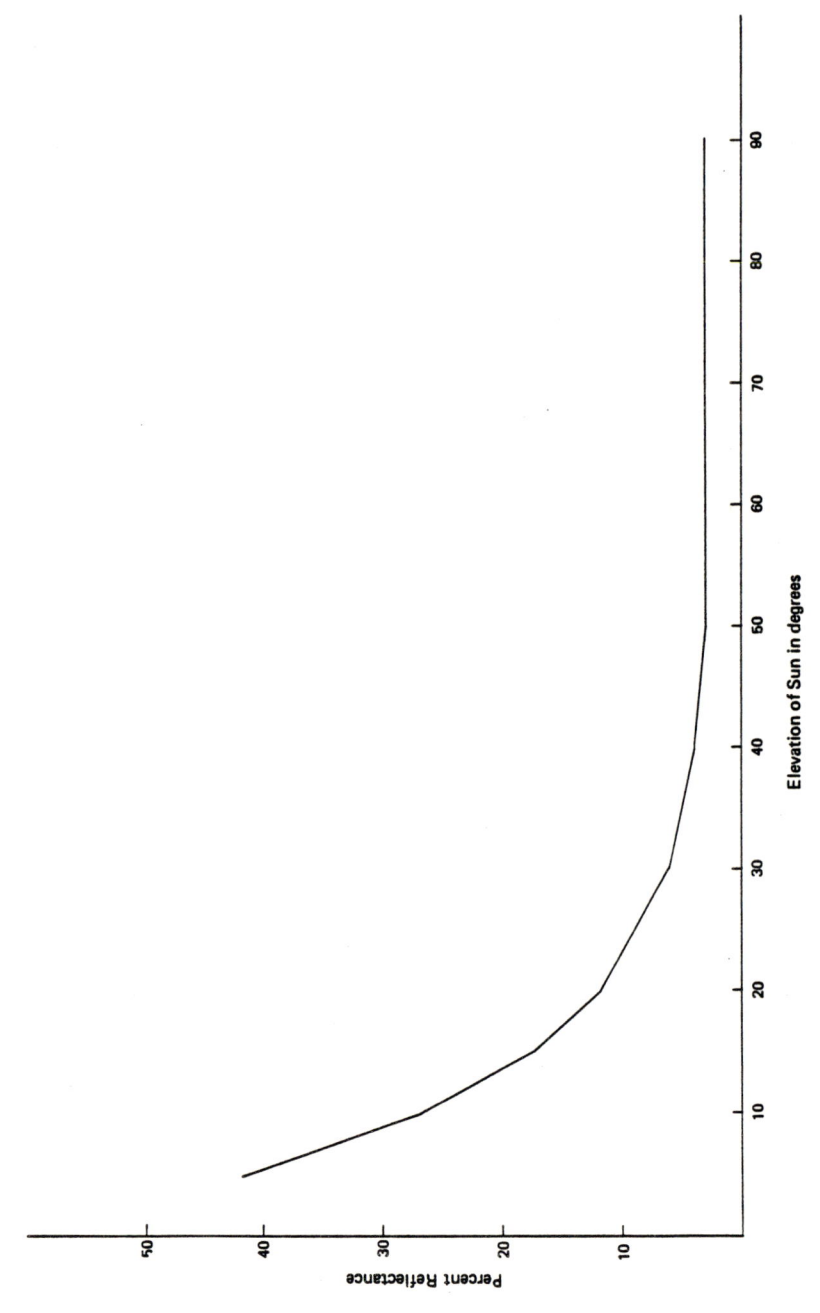

Figure 1-3. Effect of sun's elevation on reflection of sunlight from a smooth sea surface.

Table 1-L

PERCENTAGE OF THE TOTAL ENERGY, SUN AT ZENITH, AT DIFFERENT DEPTHS IN THE CLEAREST KIND OF OCEAN WATER, WATER TYPICAL OF TROPICAL AND SUBTROPICAL SEAS, AND TEMPERATE ZONE OCEANS RESPECTIVELY (AFTER HOLMES, 1957).

Depth, m	Clearest	Oceanic Water, % Tropical & Subtropical	Temperate
10	22	15	7
20	15	6	2
40	8	1.5	0.15
60	3.5	0.4	0.01

Fischer et al. (1977) made a comparison of the size distribution of suspended particulate matter and their optical properties. Fischer (1977) also described the distribution of suspended particulate matter in the equatorial region of the Pacific Ocean.

Particulate Settling

The settling of oceanic particulate aggregates, 5-500m diameter, does not conform to Stokes Law. The behavior of these aggregates is modified by electrolytes in solution, dissolved organic substances, and naturally occurring surface coatings (Chase, 1979). These factors, in combination, bring about a reduction in overall nonlinear skin friction exhibited by $\leq$ 1 order of magnitude increase in velocity over expected value for an aggregate with mean diameter of 28.5m.

Gardner (1977) has pointed out that for small particles in a column of ocean water, that is still, the velocity of settling is best portrayed by Stoke's Law. If one measures both particle size and density (this is most difficult) it is possible to estimate the velocity of fall of the particulate matter. Gardner has presented a chart in which are illustrated empirically ascertained velocities of fall for an assortment of biological particulates. The settling velocities, in cm/sec, as against particle diameters, in mm, are included as linear relationships for densities of 1.1, 1.5, 2.65 respectively at 5°C and 2.65 at 20°C.

Particle Settling in Fresh Water

Studies made on the Missouri River relating suspended sediment content to depth in the water column show a great concentration of heavy particles near the bottom (Table 1-II). Suspended clay and silt are fairly equally distributed with depth.

Table 1-II.

VARIATIONS OF SUSPENDED SOLIDS WITH DEPTH IN THE MISSOURI RIVER, NEAR KANSAS CITY. VALUES IN 100 PPM.

Height above bottom, feet	Clay	Silt	Fine	Sand Medium	Coarse
Surface	100	200	200	50	25
11	100	200	200	50	25
10	100	200	200	50	25
9	100	200	210	50	25
8	100	200	250	50	25
7	100	200	265	50	25
6	100	200	300	75	25
5	100	200	350	100	50
4	100	200	400	150	75
3	100	200	500	200	100
2	100	210	600	300	200
1	100	250	750	360	300
bottom	100	300	900	450	500

Some Basic Considerations

In the discussion of turbidity it might be useful to recall some of the fundamental rules of physics that determine the optical properties of sea water. In general when we consider turbidity we are concerned with that portion of the electromagnetic spectrum that is visible to the human eye. That region of the spectrum reaches from a wavelength of about 0.40×10^{-6} to 0.70×10^{-6} meter. It is delimited by the ultraviolet region at one end, 10^{-8} to 0.40×10^{-6}, and the infrared region at the other, 0.70×10^{-6} to 3×10^{-3}. Thus the visible light makes up an extremely narrow band in the overall electromagnetic spectrum, a region from about wavelength 400 nanometers to about 700 nanometers. Within that narrow range the light varies in color approximately as follows:

Visible Color	Wavelength, nanometers
Violet	400-420
Blue	460-480
Green	525-545
Yellow	565-585
Orange	590-610
Red	650-670

(Optical Society of America, 1966)

Measurements of turbidity, especially those involving remote sensing, such as from aircraft or satellites, may be interfered with by the phenomenon of reflectance from direct sunlight off the surface. Reflectance is defined as the ratio of the reflected radiant flux to the incident radiant flux. Reflectance varies with zenith distance (which is 0° when the sun is precisely overhead). The zenith distance of the sun is the angle formed between the zenith and the center of the disc of the sun. For example the reflectance is 2.0 - 2.4 percent when the zenith distance is between 0° and 40°; the reflectance is almost 6 percent when the zenith distance is 60°; about 13 percent at 70°; nearly 35 percent at 80°. With the sun on the horizon (zenith distance 90°) 100 percent reflectance obtains.

When light actually gets into the ocean some of it is absorbed in passing through the water column; some of it is scattered by foreign particles in the water. A useful term used in talking about absorption of light in passing through a column of water is the "characteristic absorption length" which is given the symbol called Libra. It is equal to the length of water column at which the transmittance of light equals $1/e$. Assume that a column of ocean water has a characteristic attenuation length of 10 meters; this means that light in passing through that 10-meter column of water loses all but about 37 percent of the incident energy. The attenuation coefficient in this case would have a value of 0.1 per meter. Visible light, in general, has the greatest attenuation length of the electromagnetic spectrum. For blue light (470 nanometers) in distilled water the characteristic attenuation length is about 225 meters. It is interesting to note that "as far as absorption properties are concerned, distilled water and seawater appear to be identical" (Williams, 1970).

Trace Elements

Hunter (1980) has evaluated the processes that apparently affect particulate trace material in the sea surface microlayer (Hunter, 1980). Iron and manganese decrease significantly in the microlayer; this results from the gravitational effect on mineral particles that change iron and manganese. On the other hand in the microlayer copper, zinc and lead are enriched; a condition that "probably results from flotation of particles attached to rising bubbles."

Duinker and coworkers (1979) have shown that the concentration of several metals in coastal suspended particulate matter depends to a great degree on the amount of particulate matter in each liter of seawater. There seems to be little dependence of concentration on size or particle density.

In the Thames Estuary the trace metal composition of suspended particulate matter is not the same as that of the sediment (Nelson, 1979).

Dams

Ward (1974) has shown that in the South Platte River, Colorado, the total suspended matter in the river varies as follows:

Location	Mean in mg/l
Damsite	2.2
2.4 km below dam	4.2
5 km below dam	7.4
8.5 km below dam	25.5

It is clear that the total suspended matter increases rapidly, from a low mean value immediately below the reservoir, as one moves downstream. During the study period some measurements were made during a spring flood. If those values are excluded, the mean suspended matter at the four collecting points varies from 1.9-2.6 mg/1. Ward (1974) contends that "the effect of the reservoir in clarifying the stream below extends for a considerable distance except when bankfull level is exceeded." In general suspended matter reflected major, but not necessarily minor changes in river flow.

The water clarifying effect of reservoirs as traps for sediment is well recognized. Simons (1979) has concluded that, "large reservoirs will trap almost one hundred percent of the inflowing sediment." However, small reservoirs tend to trap lesser fractions of the total silt load.

The behavior of suspended matter in freshwaters must be known in significant detail if dams, diversion structures and "river-training works" are to be constructed in a rational, nondestructive manner. Simons (1979) has outlined the elements of an adequate river analysis in this regard.

Turbidity in streams as a rule seems to be decreased by reservoirs; indeed turbidity may be comparatively low in a controlled stream. Such an effect will increase light penetration into the affected waters; an increased standing crop results. Submerged bottom plants are favored by the modification (Ward and Stanford, 1979).

Chapter 2

SOME ECOLOGICAL EFFECTS

Size of Material in the Ocean

SHELDON et al. (1972) made a general study of size distribution of particles in the ocean. They submitted a theory of the distribution of particles in the volumes from bacteria to whales. Their general conclusion was: ". . . roughly equal concentrations of material occur at all particle sizes within the range from 1 to about 10^6, i.e. from bacteria to whales." In other words, in the ocean it is their view that the same amount of material is distributed for a given volume. That material may be distributed in the form of very small particles or one large particle such as a whale; but the overall mass distribution is roughly equivalent for all volumes in the ocean.

Vertical Fluxes

The area of the ocean bottom that is 2 kilometers or less deep is not a very large percentage of the overall oceanic bottom area. Nevertheless, that rather restricted area "accounts for 85 percent of the total oceanic benthic oxygen consumption, which reflects the primary productivity of the surface waters." (Hinga, K.R. and others, 1979).

The flow of nutrient material across the interface, sediment: water was measured *in situ* in the North Atlantic with the following results:

Material	*Flux*
Total Particulate Matter	205-280 mg/m²/day
Organic Carbon	14-30 mg/m²/day

Vertical fluxes of foraminifera, radiolaria, pine pollen, metal spheres, and diatoms varied in numbers of discrete particles from as high as 700,000 diatoms/m^2/day to as little as 6-560 metal spheres (av. 65m diameter)/m^2/day.

The flux data obtained compared with the measured benthic oxygen consumption shows that the vertical flow of particulate organic carbon is enough to supply fuel for the benthos at 3500m deep; additional organic carbon must be supplied to answer the fuel requirements of benthos at 600m and 1300m respectively. "The transport of organic matter by vertically migrating organisms is suggested as the dominant additional input at the shallower depths" (Hinga and others, 1979).

Migrations and Particulate Matter

Wiebe and others (1979) have described in *Salpa aspersa* a diel vertical migration of at least 800 meters; 85 to 90 percent of the total mass of zooplankton left the top 500 meters of water during the day, in the Western Slope water of the North Atlantic Ocean.

This unusual migration has a significant effect on the flux of particulate organic matter into the sea depths. "While salp blooms may occur infrequently, both the pulsed nature and the magnitude of the transfer as opposed to a more continuous 'rain' of particles makes the event of particular importance to the ecology of deep-sea communities." (Wiebe and others, 1979). The authors postulate that important amounts of other materials, e.g. trace metals, organic compounds, organophilic pollutants, may be similarly translocated.

Deposited Material

The physical properties of suspended particulate matter will determine to a large extent the physical properties of the underlying sediment. Changes in the physical properties of the latter influence the occurrence as well as the distribution of marine-suspended particulate feeders; these in turn exert significant effects on the deposited material (Rhoads, 1974). "In the marine realm, the net effects of the activities of deposit-feeders has usually been to exclude most species of suspension-feeders"

(Tevesz, Soster, and McCall, 1980). This well-documented phenomenon has been called trophic group amensalism (Rhoads and Young, 1971; Aller and Dodge, 1974).

In freshwaters the interactions of deposit-feeders and suspension-feeders is obscure (Tevesz, Soster, and McCall, 1980).

Corals

Loya (1976) studied the effects of suspended particulate matter on a community of corals off the west coast of Puerto Rico. According to Loya: "Water turbidity and sedimentation seem to be the major factors that dictate the distribution of corals in the different reef zones." For example, in the areas known as the upper East Reef, coral diversity and living cover were found to be high; in that same area the mean turbidity of the water was low (1.5 Formazin Turbidity Units), as was sedimentation (30 milligrams per square centimeter per day). In the area called the West Reef, coral diversity and living cover were slight; mean level of water turbidity was high (5.5 Formazin Turbidity Units), and so was sedimentation (14 milligrams per square centimeter per day). The author views the differences in community structure between the two locations as attributable to turbidity and sedimentation effects. There seems to be some evidence that certain species of coral thrive better than do others in regions of heavy turbidity and sedimentation (*Montastrea cavernosa* is one such species).

Freshwater Suspended Solids (Sorensen et al., 1977)

Eroded soils produce the most important type of suspended solids on a large scale. "Sand, silt, and clay are dislodged by rainfall and overland flow and carried into streams and lakes from rural and agricultural areas, forests, and urban areas. Sediment resuspended in the course of the stream (bed load) is also an important type of suspended solids..."

"Organic suspended particulates compose an important part of suspended solids in most natural waters. Natural detrital material can be dislodged from the soil surface and enter a stream or lake. . . . 50 percent of the suspended solids being exported from the undisturbed area at Hubbard Brook were organic in nature. Often the less dense organic fraction of soil will be preferentially removed in runoff causing the organic fractions of the suspended solids to actually be enriched. This organic fraction is often higher in nutrients than the inorganic fraction of the soil. The suspended solids washed from feedlots are primarily organic material. Much of the suspended matter in urban runoff is organic. . . ."

Dredging

There is a growing understanding of the production of turbidity by human activities in the ocean, such as dredging (Nakai, 1978).

It is well-known that dredging and the disposal of spoil from the dredging operations can cause a severe problem with respect to excessive turbidity in the surrounding waters. Morton (1977) has reviewed the literature dealing with the ecological effects of the dredging in various kinds of water. His review contains useful material for evaluating on a broad scale the effect of man-made heavy turbidity.

In the process of dredging and beach mining, bottoms are disturbed to a considerable depth. Contours of the bottom or along the beach are modified and finely divided material is released and will be carried away from the operating location by tides and currents. Turbidity will be increased. Local transportation of sediment may well be modified. The bottom to a greater or lesser distance from the operating site will be blanketed with a finely divided particulate material.

"Very commonly fine sand and silt is washed out of dredge material during its passage from the bottom to the hold of the dredger and this spreads the effects more widely since surface currents are often stronger than those near the bottom." Despite this possible environmental hazard experimental and observational data, especially in the San Francisco Bay area, have demonstrated the "physical effects of dredging operations to be minor when compared to turbidity and suspended solids increases caused by natural events such as high runoff from overland or wind and wave action" (GESAMP, 1975).

Some adverse local effects of dredging may be brought about by changes in behavior of marine organisms induced by turbid clouds or by disturbances of the bottoms. One anticipates that such behavioral changes would not cause a loss of fish food organism stock but would, in all probability, change the pattern of distribution of these organisms. Fishing success might be modified and not with respect to the overall potential but rather with respect to success in a familiar area that now has been modified. Sedentary species of shellfish are particularly vulnerable to profound changes as a result of the settling out in large quantities of suspended particulate matter.

Water Quality

Iwamota and coworkers (1978) reviewed the relationship of sediment as it might be connected to water quality. The review is primarily concerned with the quality of water from the point of view of pollution. However, reference is made to many publications that are useful in interpreting the effect of turbidity and suspended solid materials on fish life.

Another example of information that is available with respect to suspended solids and the effect on fish is the publication by Sorenson et al., (1977) who review specifically publications dealing with suspended and dissolved solids as they effect a wide variety of freshwater organisms. Again, this particular publication emphasizes the fact that most experimental and observational work on turbidity effects seems to be based on freshwater studies. This does not mean that there are no marine organisms studied from the point of view of turbidity and sedimentation. It merely means that with respect to the volume of data available most of the information comes from work on freshwater organisms.

With respect to estuaries, Wellershaus (1981) cautions that dredging lets loose chemical and biological processes that do not happen in undisturbed estuaries. There are kinds of suction dredges that selectively pull up the heavier particles (size of fine sand and greater). These dredges permit the finer particles to remain effectively in the turbidity maximum. "Thus natural processes are hindered by continuous maintenance dredging." (Wellershaus, 1981).

On the other hand there are strong data that suggest that in an estuary the increased turbidity (total suspended silt load) which results from natural events, e.g. storms, is about an order of magnitude greater than that produced by dredging (Bohlen et al., 1979). The latter causes disturbances that are more narrowly localized than those resulting from natural events.

A status summary of the U.S.A. Deep Ocean Mining Effects Study (DOMES) related to manganese nodules suggests that no increased mortality of zooplankton should be anticipated in the turbid plume from nodule mining activity (Burnes, 1980). It may well be that the overall adverse effects of deep sea mining will be far less than pre-mining estimates indicated, especially with respect to the impact of turbidity changes.

Effects of Turbidity on Water Supply (Sorensen et al., 1977)

"Modern water supply treatment plants are designed to remove suspended solids within the range commonly experienced in the raw water supply. Of course, as the suspended solids load that must be removed from the raw water resources of domestic water supply. . . . An excellent source of water supply, requiring only disinfection as treatment, would have a turbidity range of from 0 to 10 units. A good source of water supply, requiring usual treatment such as filtration and disinfection would have a turbidity range of 10 to 250 units. Waters with turbidities over 250 units are poor sources of water supply requiring special or auxiliary treatment and disinfection.

". . . waters of higher turbidity (30 JTU vs. 5 JTU) may be more easily coagulated and clay is sometimes added to raw water to give this effect. Surface area, charge density, and exchange capacity of clay mineral particles all have an effect on treatability." Guarding of low turbidity waters is desirable and "effort should be made to minimize sudden changes in raw water turbidity since these affect coagulation, chlorine demand, and filterability of the water. The maximum contaminant level for turbidity in finished drinking water is one turbidity unit."

Table 2-I summarizes in general the effects of suspended particles on freshwater systems.

Table 2-I.

CLASSIFICATION OF SUSPENDED SOLIDS AND THEIR PROBABLE MAJOR IMPACTS ON FRESHWATER ECOSYSTEMS.

	Biochemical, Chemical, and Physical Effects	Biological Effects*
Suspended Solids		
Clays, silts, sand	Sedimentation, erosion & abrasion, turbidity (light reduction), habitat change	Respiratory interference, habitat restriction, light limitation
Natural organic matter	Sedimentation, DO utilization	Food sources, DO effects
Wastewater organic particles	Sedimentation, DO utilization, nutrient source	DO effects, eutroph.
Toxicants sorbed to particles	All of the above	Toxicity

*Some of these effects are a result of direct impacts of pollutant (primary effect) and some are a result of changes due to biochemical, chemical, or physical changes (secondary) or biological interactions (tertiary effects).

Chapter 3

BIOLOGICAL EFFECTS

Overview

THE exhaustive and brilliant review by Moore (1977) on inorganic particulates suspended in the world ocean includes an evaluation of the effects of these particulates on marine animals. The pertinent literature on the subject is covered thoroughly and clearly through the year 1976. This review should be consulted for definitions of terms used in discussions of turbidity and suspended particulate matter in the sea.

Several important generalizations derive from Moore's review:

a. Inorganic suspended particulates originate (decreasing importance as a source) from suspended particulates in rivers flowing into the sea, eolian matter, colloids or precipitated material, cosmic material.

b. The final fate of all nonsoluble particulates in suspension in the ocean is burial in the bottom sediment. The route to that ultimate fate may be long, and circuitous, and in part obscure.

c. The dry weight of seston (the *total* particulate matter suspended in the ocean, biotic as well as abiotic) in the ocean varies widely from place to place and time to time: 27,000mg per 1 to 0.028mg per 1 (Moore, 1977, Table II).

d. The settling out of particulates is a complex process involving salinity, particle diameter, particle surface characteristics, water current velocities, and unique physicochemical phenomena, such as the formation of flocculated aggregates.

e. Suspended particulates modify chemical and physical characteristics of the water masses involved. The modification of light penetration is most obvious; but temperature modifications, influences on electrolyte balance, the scrubbing out of trace metals from sea water, etc. by particulates suspended in the ocean must not be neglected. ". . . the concept of a simple, inert, inorganic suspension in nature is naive" (Moore, 1977).

f. Experimental studies on the physiological effects of suspended particulates on marine animals are plagued by inadequacies of available testing technology; some of these problems seem to be insuperable.

Digest of Physiological Effects (Moore, 1977)

Protozoa. The effects of turbidity on this group of marine animals are variable. Some species seem to be harmed. Others such as amoebas are unaffected. Species-by-species testing seems necessary.

Porifera. In general turbidity is harmful to sponges. Most require clean, clear water in which to thrive.

Coelenterata. Effects are variable. Some forms seem to be most tolerant of turbidity. The process of sedimentation has been proposed as the key adverse factor rather than turbidity itself. The effects on corals vary from one species to another.

Ctenophora. Mechanical injury to various tissues seems to be the main threat posed by inorganic particulate matter to this group.

Polychaeta. Low levels of turbidity may stimulate certain biological processes in this group; indeed low concentrations may be considered favorable for some species. Higher concentrations are harmful; high enough concentrations are fatal. Some species seem to thrive in turbid waters; others require very clean water.

Crustacea. According to Moore virtually no experimental information is available for this group.

Mollusca. Bivalves select particles that are passed through the filter feeding system. Some species show a change of filtration rate with change in concentration of suspended particulates. Generally excessive turbidity is harmful and may be fatal. Information on the Gastropoda and the Cephalopoda is scanty.

Echinodermata. Crinoids seem to benefit by turbid waters. Certain starfish stop feeding if the water becomes mildly turbid. Sea urchins have been eradicated by heavy turbidity caused by human activities.

Lower *Chordata.* Little information is available.

Fish. Most work on turbidity in fish has been done with fresh water species (Wilber, 1971; Moore, 1977). A concentration of suspended particulate matter of 25mg per 1 or less seems to be safe generally for fish. There may be various classes of fish with respect to turbidity tolerance. Most of the harm exerted on fish is probably mechanical. It has been suggested that the deleterious action of turbidity on oceanic fish is exaggerated.

Marine Mammals. No obvious harmful effects of turbidity are documented for this group. Interference with vision seems not to be a critical factor, because totally blind marine animals have been shown to thrive and grow fat despite that disability.

Man. The primary difficulty caused by turbidity for man in the sea is the modification of vision in turbid waters. Accuity, colors, and other aspects of human vision are adversely affected in turbid waters.

Table 3-I summarizes the effects of suspended solids on large freshwater invertebrates.

Some Specific Effects on Aquatic Organisms

Iwamoto et al. (1978) have given, in general, the adverse effects that one can expect on aquatic organisms from excessive turbidity. These include the clogging and damage to gills or other respiratory surfaces and the adherence of these materials to the chorion of eggs. Abrasive and other kinds of physical action provide ideal conditions for the entry of disease-causing organisms and the persistence of these organisms; behavioral changes are induced; and if sedimentation results to any great extent, then various life stages of developing marine organisms can become entombed. The water chemistry can be modified by adsorption or desorption, or a combination of the two. With respect to this alternative action other data indicate that adsorption of metals, for example, in the mouth of an estuary, may account for the decreased concentration of toxic metals in the open ocean as compared with such concentrations in the middle of an estuary. Turbidity can reduce photosynthesis and primary production. Material falling out from the suspended solid complex may modify the permeability in the gravel layer and may also have an effect on dissolved oxygen. Finally, the turbidity can modify vision in fish and may effect fishing on the commercial and at the level of sport fishing.

In general, according to Iwamoto and others (1978) and Muncy et al., (1979) observational studies in the field have been confirmed by laboratory studies that show that in general advanced life stages of the majority of fish species seem to be quite tolerant to any direct adverse affect of suspended solids. It should be emphasized again that the most severe impact on any fish species is probably the impairment of some stage or stages in the reproductive process. These would include adult maturation and reproductive behavior, egg and larval growth, development, and survival. "The most serious impact of sediment on salmonids occurs from sedimentation of the gravel used for spawning..." (Muncy, 1979).

Feeding

Poulet (1973) made a study of the feeding of *Pseudocalanus minutus* on natural particulate matter in the ocean. He found that this species consumed particulate matter in the range of 4-100 diameter. The species seemed to prefer food falling in the size range of 25.4-57.0 diameter. In general, particles having a diameter less than 39 were eaten more readily than were larger particles. The species has reasonable flexibility with respect to food intake. "*Pseudocalanus* was able to shift its grazing pressure from small to large particles to compensate for a reduction in density of small particles."

Studies on the fresh water shrimp, *Mysis relicta* have indicated that increased turbidity up to 1,558 (American Public Health Association turbidity units of glacial flour sediments) is not accompanied by increased mortality during an eight-day exposure period (Gregg and Bergersen, 1980). There is some suggestion that survival is better at the highest turbidity levels.

Biological Effects

Among the biological effects exerted on marine organisms by suspended particulates is a modification of the rate of pumping water through the body.

Gerrodette and Flechsig (1979) measured the pumping rate of *Verongia lacunosa*, an oceanic sponge found in the tropics, as modified by suspensions of clay in the surrounding water. In clear water the sponge pumped about 1-6 liters of water per hour or a total volume of about half a liter. If clay is added to the water to give a concentration of 11 milligrams clay particulates per liter of water or more, the pumping rate is reduced. At a concentration of 3 milligrams clay particulates per liter, no change in pumping rate was observed.

Sponges exposed to 95 milligrams clay particulates per liter of ambient water for four days responded by "a continuing decline in the pumping rate." The sensitivity of these and related sponges may limit their distribution.

The results indicate that the pumping rate drops inversely to the concentration of the suspended particulate matter. There is no indication of a sudden shut down of pumping at a critical threshold concentration of suspended clay particulate. This relationship may be expressed in a general way as follows:

$$y = -0.11 - 0.01x; r = -0.92; r^2 = 0.84$$

where y is the difference between pumping rate of the control (clear water) and the turbid water sponges expressed in liters per hour and x is the turbidity of the ambient water expressed as milligrams of suspended clay particulate matter per liter of water.

Feeding Efficiency

The efficiency of feeding in some bivalves is related to the presence of eu-latero-frontal cirri on the gill filaments. These structures act as screens with mesh size about 0.6 × 2.7m (Møhlenberg and Riisgård, 1978). Eleven species of bivalves that have these structures retain completely particles pumped through the body that have a diameter of 4m or greater. In these same species "Particles down to 1m were in most cases efficiently retained." In two species of *Pecten* that do not have eu-latero-frontal cilia the efficiency of particle retention dropped off as particle diameter fell below 7m. With particles of 1m diameter only about 20 percent of the material filtered was retained. Møhlenberg and Riisgård (1978) conclude that "the efficiency of the bivalve gill in retaining particles probably depends upon the integrated activity of all the ciliary systems of the gill."

Feeding Selection

An imaginative model has been presented to predict the selection of particles by size in deposit feeders. Factors such as gut passage time, assimilation efficiency, gut volume, costs of particle rejection, and selection of particle size are interrelated in the model (Taghon, Self, and Jumars, 1978). The authors are now testing and validating their model.

The model derives from similar studies on filter feeders by Lehman (1976) and Lam and Frost (1976).

Feeding Rates

The feeding rates of *Mytilus edulis* from the Wadden Sea and from "inner Danish marine waters" were measured by Theisen (1977). Dyed yeast served as food. Different concentrations of talcum were suspended in the ambient water to simulate variation in the natural amounts of suspended particulate matter. The rate of feeding decreased sharply as the concentration of talc in the water surrounding the mussels increased. Mussels from different locations responded differently to high levels of suspended talc. "At high concentrations of talcum *Mytilus* from the Wadden Sea fed at about twice the rate as *Mytilus* from other Danish waters."

If relative feeding rates are used for comparison (setting 36.4mg/l concentration of talcum as 100 on the relative rate scale), the following relationships may be calculated:

For inner Danish mussels,
$$y = 152 - 43.3 \log x; r = -0.98$$
For Wadden Sea mussels,
$$y = 161.6 - 39.6 \log x; r = 1.0$$
where y is the relative feeding rate and x is the concentration of talcum in mg/l.

Theisen (1977) contends that the relative rates of feeding in the two groups of mussels, especially at higher levels of suspended particulates, are correlated with the size difference in labial palps: mussels from the Wadden Sea have palps about two times the size of those from other Danish waters. The larger palps are seen as morphological adaptations to the comparatively turbid waters of the Wadden Sea. The following generalization is made: "Bivalves living in waters of high turbidity generally have larger palps than those living at low turbidity."

Filter Feeder Feeding

It is not always easy to ascertain whether polluting chemicals or other materials are taken up by marine filter-feeding animals by way of suspended particles, in the food, by direct absorption through the body surface, or some other route. In *Mytilus*, there is good evidence that certain metals (zinc, manganese, cadmium, and the metalloid selenium) are taken up primarily from suspended particulate matter (Bryan, 1979).

In *Mytilus edulis*, particulate material is taken up not only by the digestive system but also to a significant extent by way of the gills; pinocytosis transfers the particulates to other tissues through the activity of circulating amebocytes. The process is energy dependent.

Although experimental evidence to show the importance of particulate matter as a source of organic pollutants in marine animals is not available, what is available suggests that particulates are very important: e.g. the uptake, by the copepod *Calanus helgolandicus*, of the hydrocarbon naphthalene originating from petroleum depends heavily on particulate food taken into the digestive system.

The feeding rate of a large number of filter feeding marine animals seems to be modified by the concentration of suspended particulate matter surrounding these animals and also the size of the suspended particles. For example, the rate at which the tropical marine sponge *Verongia lacunosa* pumps water varies from 1 to 6 liters of seawater per hour if the sponges are kept in clear seawater. If these sponges are put in seawater, which carries in suspension clay particles, the rate of pumping water varies depending on the concentration of the suspended clay. If the concentration is 11 milligrams clay per liter or greater, the pumping rate is decreased significantly. If the concentrations of clay are no more than three milligrams per liter, no change in the rate of pumping is recorded. Some sponges of this species were put in seawater containing suspension of clay particles for a period of four days. If the concentration was 95 milligrams clay per liter, there was a steady decline in the rate of pumping water (Gerodette and Flechsig, 1979).

An interesting series of experiments was made by Fenchell (1980) in which latex beads of different diameters were fed to fourteen species respectively of ciliated protozoa. These ciliates are what are known as bacteria-eating holotrichs. Each species apparently has a preference for a distinct spectrum of particle sizes that are retained in the feeding apparatus and eventually ingested and digested if the material is digestible. The ciliates studied for the most part retain suspended particles down to a diameter of 0.2 micrometer in diameter; one species actually retains particles that have a diameter as small as 0.1 micrometer. The most efficient retention is observed in these protozoa when the particle size is somewhere between 0.3 and 1 micrometer. An interesting point is made that data from the literature and the work of Fenchel (1980) give no support to the view that suspension feeding ciliates can discriminate qualitatively among different types of particles. The data available strongly demonstrate that these unicellular animals can discriminate with respect to size of the particles taken in.

According to Fenchel (1980b) ciliated protozoa that feed on particles larger than one micrometer in diameter resemble metazoan organisms that feed on suspended particles with specific reference to the capacity for concentrating particulate matter obtained from dilute suspensions. It is therefore understandable that ciliated protozoa, in general, are found where plankton are found; they play a significant role in harvesting phytoplankton for food. In the open ocean and in the open waters of large lakes, bacteria are found in comparatively low numbers; therefore ciliates cannot thrive under these conditions. In certain parts of marine or fresh water bodies where pollution has raised the bacterial count significantly, ciliated protozoa may be found in reasonable numbers.

Fenchel (1980b) gives the ranges of water volume cleared per millimeter of oxygen consumed for several groups of suspension feeding metazoa. The values for different groups in millimeters per liter are as follows: sponges, 4-23; bryozoa, 12-60; echinoderm larvae, 12-170; lamellibranchs, 4-79; lamellibranch larvae, 5-15; tunicates, 13-300; copepods, 5-60; Daphnia, 1.3-2.6.

In the various ciliated protozoa similar pumping rates for water in liters per milliliter of oxygen consumed are as follows: *Glaucoma*, 0.8-8, *Colpidium*, 0.3-3.8; *Paramecium*, 0.8-1.3; *Stentor*, 0.005-0.2; *Bursaria*, 0.002-0.15.

Fenchel (1980b) emphasizes in conclusion: "There can be no doubt that ciliates play a significant role as bacterial consumers in nature. In the oligotroph water masses of oceans . . . ciliates cannot be held responsible for maintaining the low concentration of bacteria characteristic for these waters." It seems that other organisms are involved. A suggested group, according to Fenchel, are the heterotrophic microflagellates.

Feeding Rate

The filtration rate in two species of molluscs, *Mytilaster lineatus* and *Cerastodermia lamarcki*, has been shown to vary inversely with the concentration of the solution being filtered (Kasymov and Likhodeyeva, 1979).

The volume of fluid filtered was calculated by a well-accepted formula (Williamsen, 1952):

$$V = m \frac{\ln C_0 - \ln C_3}{t} - a$$

in which V is the filtration speed in milliliters per hour; $C_0 - C_t$ the concentration of suspended material; m the volume of the aquarium; t is time in hours; the symbol a is a correction factor for suspension, precipitation or evaporation of oil.

It was found experimentally that the rate of filtration decreases directly with the increase of the concentration of suspended particles. For example, with a suspension of yeast cells the following results obtained with *Cerastodermia*:

Concentration of yeast cells	Rate of filtration
600	25-42
300	25-29
200	53-67

With *Mytilaster* the filtration rate is as follows:

Concentration of yeast cells	Rate of filtration
200	20-60
300	18
600	18-59

If clay particles were in suspension the filtration rates were as follows:

Concentration of clay, mg/l	Filtration rate Mytilaster	Cerastodermia
200	450	450
300	180	108
600	89-334	89-1334
700	56-58	-

The filtration rates for any suspended substance varied directly with the size of the mollusc tested. For example *Mytilaster* filtering suspended clay at a concentration of 600mg/l showed the following:

Weight of specimen, mg	Filtration rate	S
39	89	3
63	242	1
84	252	1
99.6	334	1

Cerastodermia showed the following rate-weight relations filtering clay at a concentration of 600mg per liter:

Weight, mg	Filtration rate	S
39	89	3
63	242	1
84	252	1
99.6	334	1

These molluscs also filter oil but apparently the oil is poisonous to eventual lethality for them.

Riisgård and others (1980) found that veligers of *Mytilus edulis* cleared most effectively from suspensions particles with diameters from 2.5-3.5 m. Particles smaller than 1m diameter or larger than 8 to 9m are not cleared from suspensions. Juveniles of the species, 1 to 4mm long, cleared particles having diameters of 4m or greater with 100 percent effectiveness. Particles less than 4m in diameter were successively less effectively retained; only 20 percent of particles having a diameter of 1m are cleared from suspensions by juveniles of *Mytilus edulis*. The effectiveness of particle removal from suspensions by veligers and juveniles of *Mytilus edulis* agrees with results obtained in studies of specimens of *M. edulis* that are 1.6, 4.3, and 8.6 cm. long respectively (Møhlenberg and Riisgård, 1978).

Jørgensen (1975) has studied the particle retention capacity of the gills in *Mytilus edulis* using experimental, surgical, and pharmacological approaches. He found that the gills cleared suspended particles at a high rate. Most particles down to 3-5 m were removed from suspension; about 50 percent of the 1-2 m particles were removed. The data from this study taken in conjunction with information in the literature strongly suggest that the greatly effective retention capacity of the gills in *Mytilus* for particulates results from "the integrated activity of the ciliary systems." Disruption of these integrated systems causes inadequate particulate retention.

There is, with respect to deposit feeders, continuing controversy over selectivity or no selectivity, and if the former how it is done. Self and Jumars (1978) used glass beads in an experimental study of the matter in marine polychetes, *Pseudopolydora kempi japonica, Pygospio elegans*, and an unnamed "ampharetid." They showed that the ampharetid selects for ingestion particles based on its specific gravity. All the polychetes tested selected particles for ingestion based on surface texture of the particles. Etched glass beads were ingested in preference to smooth. *Pygospio* exhibited their selectivity most strongly of the three species.

In general particles with greater specific gravity stay in the gut of all worms tested for a shorter time than those of lesser specific gravity. This fact calls into question the evaluation of bulk gut contents in experiments on selective ingestion.

The authors conclude that "particle specific gravity, particle surface texture, and other particle characteristics are likely to be important parameters in accurately evaluating the resources utilized by deposit feeders" (Self and Jumars, 1978).

Winter (1969) has made an extensive study of filtration rates in the mussels *Arctica islandica* and *Modiolus modiolus* with respect to concentration of living particulate matter as well as other variables. Fundamentally the rate varies with size of the animal according to the familiar allometric equation,

$$y = a \cdot x^b$$

where y is the filtration rate, a is the specific capacity of a mussel containing a one gram wet weight of soft parts, x is the wet weight of the soft parts of the specific mussel under consideration, and b the relationship between body size and filtration rate. The equation is written for ease of handling,

$$\log y = \log a + b \cdot \log x$$

With increasing concentrations of suspended living cells (*Clamydonas*) the filtration rate of both species decreases.

Certain mussels, e.g. *Mya arenaria*, *Cardium edule* among others, live in waters that normally have a high amount of seston; they are adapted to high levels of particulate matter; they are not capable of adjusting to low concentrations of suspended particulate food by increasing the filtration rate.

Palmer (1980) has discussed the changes in filtration rate in the bay scallop and in the oyster as a function of suspended particulate matter.

Hildreth and Mallet (1980) showed that the retention efficiency of *Mytilus edulis* for diatoms having a diameter of $5.0_\mu M$ remained constant at a level of about 97 percent over an extremely wide range of concentration of suspension density.

Applied Biological Considerations

In a review for the Environmental Protection Agency (Sorensen et al., 1977) some applied aspects of turbidity and fresh water fishing are considered: "Clear farm ponds produced from 1.7 to 5.5 times the total weight of fish in turbid ponds. Largemouth bass were most affected by turbidity. Interference with light penetration lowered plankton productivity by 8 to 12.8 times in turbid waters as opposed to clear waters. This reduction in productivity limited the amount of available food for fish. Individual channel catfish grew faster in clear ponds but greater total weights were obtained in muddy ponds due to lack of competition. The presence of carp (which increased turbidities) reduced the growth of bass and bluegills but led to increased yields of channel catfish and bluegills. A clear reservoir attracted more anglers, yielded greater returns per unit of fishing effort, as well as desirable species, and was aesthetically more attractive."

It was also reported that "the mortality of fish exposed to suspensions of wood fibers such as those from pulping plants, depended on the species of fish, type of wood fibre, processing method, dissolved oxygen concentration, and to a lesser degree, water temperature." Using young of the year of fathead minnows (*Pimephales promelas*) and walleyes (*Stizostedion vitreum vitreum*), they found that ground conifer wood was the most lethal and had the greatest effect on walleyes fingerlings, and that "ground wood pulps were more lethal than chemical pulps."

Pollution

The joint group of experts on the scientific aspects of marine pollution (GESAMP, 1975) has considered in some detail the problems of pollution of the ocean and has devoted significant concern to turbidity and suspended particulate matter. They point out that it is essential to have information about any kind of pollutant so that information with respect to whether the waste is liquid, solid, or a solid in suspension can be taken into consideration in evaluating the effects of such polluting materials. The density of the waste material also must be considered in evaluating the outcome of this waste no matter what it might be. It is known for example that "Settling velocity will be influenced by the shape, size and density of the particles, and aggregated matter will settle more rapidly than individual particles of the same density. Under stratified conditions, particulate matter may be retained or have its vertical dispersion suppressed in a pycnocline layer."

There are several ways that suspended particulate matter can influence and modify the oceanic environment. If the particulates settle in considerable quantities over a circumscribed area the bottom living organisms usually will be adversely affected. "Although in some sea areas the natural suspended particle load is high, addition of suspended matter will increase turbidity and may cause discoloration of the water with possible adverse effects on fisheries and recreational interests. Certain forms of particulate waste may clog gill surfaces of marine fish, crustacea and bivalve shellfish. If a waste is particularly insoluble and positively bouyant, it will float and shipping or amenity interests may be adversely affected" (GESAMP, 1975).

As a general rule, the velocity at which waste particulates will settle to the bottom varies. It does seem safe, however, to use a settling rate of one meter per hour for materials in the flocculated state. If the particulate material is virtually "neutrally bouyant" it can be trapped in pycnocline layers.

Another important aspect of waste disposal in the ocean results from the fact that most chemical compounds eventually are buried in the sediments. The rate at which polluting materials move from the ocean waters into the sediment is not well understood. Moreover the residence times of specific substances in the bottom sediments is somewhat obscure. The eventual mobilization of these varied materials in the sediments is not at all well known. In order to clarify many aspects of how turbidity (taken in the broad sense) is influenced by dumping of waste material in the ocean, it seems that additional careful and expensive measurements are required on the concentration and distribution during descent after dumping of these materials. The turbidity generated should be worked out experimentally. Obviously precise measurements of separating velocity and eventual dispersion in relation to a wide variety of physical factors seems to be needed in order to make intelligent recommendations about the sites, rates, and other conditions of marine waste disposal.

Ocean Mining

The mining of ocean bottoms probably will bring around

only negligible effects: "Most of these effects would be quite local and comparable to those produced by dredging but if the exploited material was a metal ore, or a metalloid, then some potentially dangerous fine material might accumulate in the bottom deposits and be subsequently released, and some soluble material might enter the water column directly" (GESAMP,1975).

Experimental Fish Development

In an attempt to ascertain the effects of turbidity, caused by marine mining, on fish development artificially fertilized herring eggs (species *Clupea harengus*) were exposed to various steady concentrations of suspended silt in the range of 5 to 300 mg/l. Short-term exposures to 500mg suspended silt per liter were also tested. Various developmental stages of the herring were used as test organisms. Embryonic development was not influenced in any of the test suspensions used. The conclusion deriving from this experimental work is that no harmful effects need be anticipated to herring development from turbidity that may accompany dredging, mining, and similar activities in the ocean (Kiorboe et al. 1981).

Near Shore Situation

With respect to the near shore region exploited in the ocean there seems no question that the increase in turbidity from such activity will be considerable. Finely grained particulate matter will be carried over large distances along the shore and out into the open ocean. "The possible implications of dredging for the material budget of the beach zone should always be considered" (GESAMP, 1975). Because dredging can significantly modify the topographic situation locally such an operation can severely change off-shore suspended material transport patterns. Along the shore these patterns depend primarily on wave conditions.

In off-shore dredging, currents, and other mixing conditions seem to be generally adequate so that specific concentrations of waste material will be reduced to inconsequential levels within five to ten hours.

Off-Shore Sand and Gravel

Kullenberg (1975) has evaluated the turbidity problems associated with the dredging of off-shore sand and gravel. He points out that the rate of dispersion in potential dredging areas will be fairly high, therefore, increases of turbidity will be relatively small. He estimates the increase of turbidity as follows: "a large dredger may recover say $10^4 m^3$ in five hours. Assume as a working basis that 1 percent of the recovered material is lost with the overflow in the form of fine-grained material with a density of $2.5 g/cm^3$. Thus, we dispose of 50 tons per hour in the water, corresponding to $2.5 \times 10^4 mg/1$ overflow."

From the studies that have been made, it seems apparent that the actual operation of dredging as well as its frequency can be modified in order to insure that turbidity at any given time will produce no adverse effects on the living organisms in the ocean.

Deep Sea Mining

Information concerning the mining of the deep sea ocean bed is limited. Most experiments that have been made have local value only and may not be extrapolated without severe precautions. It seems that the first distribution of resuspended material brought about by deep sea ocean bed mining depends particularly on the technique of mining being used. "Taking the manganese nodule mining as an example, and assuming that as much sedimentary material as the amount of nodules is brought to the surface and released with the mining effluent into the surface layer, an estimate of the turbidity increase in a single 24-hours operation suggests that the initial concentration will be of the order of 100-1000 times the natural (Ambient) concentration in the top 50-100 meters. Over a period of 10 hours this concentration will normally decrease by at least a factor of 100" (GESAMP, 1975).

Kullenberg (1975) addressed himself to potential turbidity problems associated with deep sea mining. One assumes that the amount of suspended material that will be obtained will equal the amount of manganese nodules that are taken in deep

Biological Effects

Table 3-I.
SUMMARY OF SUSPENDED SOLIDS EFFECTS ON AQUATIC MACROINVERTEBRATES (Sorensen and others, 1977).

Organism(s)	Effect	Suspended Solid Concentration	Source of Suspended Solids	Comment
Mixed Populations	Lower summer populations		Mining area	
Mixed Populations	Reduced populations to 25%	261-390 ppm (Turbidity)	Log dragging	
Mixed Populations	Densities 11% of normal	1000-6000 ppm		Normal populations at 60 ppm
Mixed Populations	No organisms in the zone of settling	>5000 ppm	Glass manufacturing	Effect noted 13 miles downstream
Chironomus & Tubificidae	Normal fauna replaced by (Species Selection)		Colliery	Reduction in light reduced submerged plants
Cheumatopsyche (Net spinners)	Number reduced	(High concentrations)	Limestone Quarry	Suspended solids as high as 250 mg/l
Tricorythoides	Number increased		Limestone Quarry	Due to preference for mud or silt
Mixed Populations	90% increase in drift	80 mg/l	Limestone Quarry	
Mixed Populations	Reduction in numbers	40-200 JTU	Manganese Strip mine	Also caused changes in density and diversity
Chironomidae	Increased drift with suspended sediment		Experimental sediment addition	
Ephemoptera, Simuliidae, Hydracarina	Inconsistent drift response to added sediment		Experimental sediment addition	

sea mining. Assume that the amount of nodules taken is equal to 10kg/m². Therefore, a minimum surface area of $5 \times 10^4 m^2$ or roughly a square measuring 220 meters on each side would be covered by the mining operation in 24 hours. "Assume a surface layer current of the order 12cm/sec or 10^4 m/day and assume further that the overflow becomes evenly distributed in the top 100m. Then the dilution volume is of the order $2 \times 10^8 m^3$, and the concentration of suspended matter (assuming that all goes into suspension) 2.5mg/l. This is a conservative estimate in the sense that most likely the value is less since not all goes into suspension, the currents can be stronger, . . . After ten hours the maximum concentration is expected to be in the range of 0.05-0.01 mg/l." Compared with the natural concentration of suspended matter in oceanic surface layers this calculated amount is not too far from what one observes naturally to occur.

Biological Effects of Turbidity on Fish

Laboratory and field studies have been reviewed by Muncy et al. (1979) to ascertain what is known about the effects of suspended solids and sediments on reproductive success in warm water fish. There are a number of controversial aspects of the problem. Moreover, it seems that there are variations from species to species with respect to the effects of suspended solids on reproductive success. Variations in year-class strength of important fishes are not as yet correlated with sediment loading, concentrations of suspended solids, or sedimentation rates. There was a renewed interest in suspended solid effects on aquatic ecosystems especially during the 1970s. During that decade the published literature and symposia reported numerous laboratory bioassays and ecological field studies on this general topic. Species and stages of warm water fishes are not equally susceptible to suspended solids. Only limited circumstantial evidence is available on the potential effects of turbidity on gonad development in fish. There seems to be substantial evidence that reproductive behavior may be affected in various ways by suspended solids and sediment, specifically with respect to spawning time, place of spawning, and spawning behavior. The more adaptively successful species of fish apparently do not carry out reproductive

Table 3-II.

SOME EFFECTS OF TURBIDITY ON SELECTED FISH SPECIES
(Sorensen and others, 1977).*

Species	Turbidity at First Adverse Reaction	Turbidity at First Death
Golden Shinner *(Notemigonus crysoleucas)*	20-50,000 ppm	50-100,000 ppm
Mosquitofish *(Gambusia affins)*	40,000	80-150,000
Goldfish *(Carassius auratus)*	20,000	90-120,000
Carp *(Cyrinus carpio)*	20,000	175-250,000
Red Shinner *(Notropis lutrensis)*	100,000	175-190,000
Largemouth Black Bass *(Micropterus salmoides)*	20,000	101,000 (average)

* "Clear farm ponds produced from 1.7 to 5.5 times the total weight of fish in turbid ponds. Largemouth bass were most affected by turbidity. Interference with light penetration lowered plankton productivity by 8 to 12.8 times in turbid waters as opposed to clear waters. This reduction in productivity limited the amount of available food for fish. Individual channel catfish grew faster in clear ponds but greater total weights were obtained in muddy ponds due to lack of competition. The presence of carp (which increased turbidities) reduced the growth of bass and bluegills, but led to increased yields of channel catfish and bluegills. A clear reservoir attracted more anglers, yielded greater returns per unit of fishing effort, as well as desirable species, and was aesthetically more attractive" (Sorensen et al., 1977).

activities during the time when turbidity of the water is highest. Fishes with more complex patterns of reproductive behavior are more vulnerable to interference of spawning and reproductive activity as a result of suspended solids. A number of critical behavioral phases in the spawning process may be adversely modified. The incubation stage is particularly susceptible to adverse effects from sediment whether suspended or deposited.

Specifically, it may be said that suspended sediments can directly affect juvenile freshwater, warm water fishes at sublethal levels by reducing site-feeding distance, disrupting activity and respiratory patterns, and changing migration and orientation responses. At levels at 2,000 parts per million, stress reactions have been observed; whereas, at levels between 69,000 to 200,000 parts per million mortalities generally have been recorded for

Table 3-III.
EFFECTS OF SUSPENDED SOLIDS ON NON-SALMONID FISH (Sorensen and others, 1977).

Fish (Species)	Effect	Concentration of Suspended Solids	Source of Suspended Materials	Comment
Mixed fish populations	Decrease in occurrence	Turbidity increase		
Mixed fish populations	Critical levels affecting populations	100-300 ppm	Industrial	England, Scotland, and Wales fisheries
Perch (Perca flavesiens)	High egg mortality	(Silting)		
European Pike Perch (Lucioperca lucioperca)	High egg mortality	(Silting)		
Zebra (Brachyolanio rerior)	Earlier egg hatch and no increase in egg mortality	18,000-30,000 ppm	Limestone dust	Fry died within 4 hours at 74,800
Barbel (Barbus fluviatilis)	Decreased migration	(Increasing turbidity)		
European eel (Anguilla anguilla)	Increased migration	(Increasing turbidity)		
Smallmouth bass (Micropterus dolomieui)	Successful nesting, spawning, hatching	(Sporadic periods of high turbidity)		

species experimentally challenged. Fishes reported to have adapted to turbid flowing waters appear to have evolved structures and sensory organs which reduce the effect of suspended solids on their bodies and improve food searching. Beneficial effects upon fishes of increasing suspended sediment levels have been attributed to escape from predation for species, such as catfish with highly developed taste and smell organs. Growth rates of such fishes can be reduced because of more juveniles feeding upon the same or reduced food base in turbid waters.

Experimental results concerned with the direct effects of suspended sediments on juvenile warm fishes have been limited mainly to high level, short-term mortality investigations. Long-term studies indicate decreased standing crops and sometimes slower growth rates of fish in ponds containing high levels of suspended sediments. However, many of the results from studies and the literature are confounded by the presence of additional fish species such as carp. Their influence on the results cannot in any way be deduced. The modes of action and exact stages of life were not experimentally demonstrated or directly implicated in all too many laboratory studies. At the present time it seems that rigorously designed laboratory and pond experiments should be designed to test the manner of action of suspended sediments on young fish in estuaries and in the open ocean.

Turbidity in the Great Lakes

Swenson (1978) has made a study of the influence of turbidity on the abundance of fish in western Lake Superior. This report contains one of the few tests of the effects of turbidity on fish. Because of the dearth of information on the biological effect of turbidity in fish, this report from a freshwater lake is included here in this study of turbidity in the marine environment. It is, I think, important also to recognize that the Great Lakes on the North American continent are by definition considered to be part of the "marine environment" legally in the United States.

Both field and laboratory studies were used in this overall research in order to measure the behavioral response of fish and the resulting changes in fish species interrelationship in western

Lake Superior. The direct effects of red clay turbidity on survival and growth of larval lake herring, *Coregonus artedii*, were recorded. Field measurements indicated that light penetration in western Lake Superior is reduced significantly even at low levels of red clay turbidity. Zooplankton and fish abundance and distribution were influenced by turbidity. Zooplankton abundance and distribution were highest near the surface in plumes of red clay turbidity. In response to turbidity smelt, *Osmerus mordax*, move into the upper 12 meters of water where their predation on larval fish increases. Predation by smelt on larval lake herring was identified as a potentially important factor contributing to the decline of the formerly abundant western Lake Superior lake herring population and the commercial fisheries that depended upon it. Walleye, *Stizostedion vitreum*, and lake trout, *Salvelinus namaycush*, demonstrated opposite responses to turbidity. Walleye concentrated in turbid water where food availability was apparently greater. Lake trout showed partial avoidance of turbidity in the lake and in laboratory turbidity gradients.

Muncy et al. (1979) in their review report for the Environmental Protection Agency discuss the effects of suspended solid materials on reproduction and early life in warm water fish. They point out that there is a limited fund of information on fish species with respect to the impacts of suspended solids on reproduction. With respect specifically to warm water species they point out: "Laboratory and field studies during the 1930-50s examined direct mortality as the result of extremely high levels of suspended solids. Controversy ensued in the 1940-60s over the impacts of turbidity on fish populations in the Great Lakes and midwestern rivers. Variations in year-class strength of important fishes have not been correlated with sediment loading, concentrations of suspended solids, nor sedimentation rates. Renewed interest in suspended solid impacts on aquatic ecosystems was evident in 1970s . . ."

In general, various fish species are not similarly susceptible to the effects of turbidity. It is also true that various stages in the life cycle of fishes are not similarly susceptible to suspended solids in the ambient water. There is some extremely limited suggestive evidence that turbidity under unique circumstances might have an effect on gonadal development in fish. There seems clear evidence that reproductive behavior is influenced by suspended solids in the water. According to Muncy et al. (1979): "Fishes with complex patterns of reproductive behavior are more vulnerable to interference by suspended solids at a number of critical behavioral phases during the spawning process. Incubation stage is particularly susceptible to adverse effects from sediment..."

From an overall evaluation of the published literature it seems safe to conclude that the larval stages of a goodly number of species of fish are less tolerant of turbidity than are the eggs or the adult stages of those fish species. The lethal levels for suspended particulate concentrations are related to a variety of other factors involving age and the particular species involved. Moreover, such factors as particle size of the suspended material, the shape of the individual particles, the concentration in milligrams per liter or some other unit, as well as the turbulence of the water being considered, all interact with the age-specific and species-specific characteristics to determine lethal levels for turbidity. "Increased suspended solids reduce sight-feeding distances, disrupt activity and respiratory patterns, and change orientation responses of some larval and juvenile [fish species]" (Muncy et al. 1979).

Some species in limited numbers have been found to circumvent in a successful manner the harmful effects of persistent high concentrations of suspended material in the environment through a combination of functional and behavioral adjustments which lead to successful survival in strongly turbulent water conditions.

"Although unequivocal experimental evidence demonstrating causal relationship between suspended solids and sediment on reproduction of... fishes was scarce, generalizations from the overwhelming body of independent observations suggested that most ... fish assemblages have been affected and species composition have been altered because of sediment effects on the more sensitive species." Moreover entire aquatic communities including plankton, macroinvertebrates, as well as fish life have been significantly modified in the face of highly turbid conditions in the ambient waters (Muncy and others, 1979).

Water Quality

Iwamota and coworkers (1978) reviewed the relationship of sediment as it might be connected to water quality. The review is primarily concerned with the quality of water from a point of view of pollution. However, references are made to many publications that are useful in interpreting the effect of turbidity and suspended solid materials on fish life.

Dredging

It is well-known that dredging and the disposal of spoil from the dredging operations can cause a severe problem with respect to excessive turbidity in the surrounding waters. Morton (1977) has reviewed the literature dealing with the ecological effects of the dredging of various kinds of water. His review contains useful material for evaluating on a broad scale the effect of man-made heavy turbidity.

Another example of information that is available with respect to suspended solids and the effect on fish is the publication by Sorenson et al. (1977) who review specifically publications dealing with suspended and dissolved solids as they affect a wide variety of freshwater organisms. Again, this particular publication emphasizes the fact that most experimental and observational work on turbidity effects seems to be based on freshwater studies. This does not mean that there are no marine organisms studied from the point of view of turbidity and sedimentation. It merely means that with respect to the volume of data available most of the information comes from work on freshwater organisms.

Iwamoto et al. (1978) has given in general the adverse effects that one can expect on aquatic organisms from excessive turbidity. These include the clogging and damage to gills or other respiratory surfaces; the adherence of these materials to the chorion of eggs. The abrasive and other kinds of physical action provides ideal conditions for the entry of disease-causing organisms and the persistence of these organisms; behavioral changes are induced; if sedimentation results to any great extent, then various life stages of developing marine organisms can become entombed; the water chemistry can be modified by adsorption or desorption, or a combination of the two. With respect to this alternative action other data indicate that adsorption of metals, for example, in the mouth of an estuary, may account for the decreased concentration of toxic metals in the open ocean as compared with such concentrations in the middle of an estuary; turbidity can reduce photosynthesis and primary production; material falling out from the suspended solid complex may modify the permeability of the gravel layer and may also have an effect on dissolved oxygen. Finally, the turbidity can modify vision in fish and may effect fishing on the commercial and at the level of sport fishing. Observational studies in the field have been confirmed by laboratory studies, which show that, in general, advanced life stages of the majority of fish species seem to be quite tolerant to any direct adverse effect of suspended solids. It should be emphasized again that the most severe impact on any fish species is probably the impairment of some stage or stages in the reproductive process. These stages would include adult maturation and reproductive behavior, egg and larval growth, development, and survival. "The most serious impact of sediment on salmonids occurs from sedimentation of the gravel used for spawning ..." (Muncy and others, 1979).

It is important to keep clearly in mind the fact that reproductive movements are influenced by turbidity in some fish species and not in others. The influence in different species is not necessarily in the same direction. For example, it now seems fairly well established that remarkably high concentrations of suspended solid particulate matter in rivers do not prevent or inhibit the migrating salmonids from reaching their goal. It is true that if a choice is available the salmonids may preferentially use a clear water route rather than a turbid water route. It is further known that the American shad and the striped bass regularly swim through highly turbid tidal waters as part of their migration into spawning areas. An interesting observation is that the movement of the European eel, *Anguilla anguilla*, seems to be aided by elevated turbidity in the water.

With respect to embryonic development several factors must be kept in mind. First of all, eggs that float seem to be resistant to suspended solids. It has been claimed, for example, that the capable larvae of the striped bass are somehow "pre-adapted" to a silt-laden water. The eggs and the larvae of the striped bass normally float and do not settle down on the bottom.

Eggs that end up on the bottom can be harmed by the settling out of suspended material. If the egg is covered up with sediment physiological damage can result to the embryo as a result of lack of oxygen or the build-up of waste materials. The damaging effect of abrasion and other physical forces on the chorion should not be ignored. Laboratory studies have shown that the eggs of certain fish species tolerate, with no increase in mortality, concentrations of suspended solids up to 1,000 milligrams per liter. Among these species are the alewife, *Alosa pseudoharengus*, the blueback herring, *Alosa aestivalis*, and the American shad.

Further there seems to be strong evidence that the eggs of the striped bass do not need to be suspended up from the bottom in order to hatch successfully. It is obvious at the same time that the eggs can be suffocated from silt; fungus infection can result from contact with a substrate that is contaminated with fungus. Moreover various unidentified factors related to water quality around the egg may result in elevated mortality if the eggs happen to end up on an unfriendly substrate. Ordinarily in nature the eggs of the striped bass are suspended in a column of water. Sediment that falls to the bottom has a limited effect on them. However, if the suspended particulate solids should be greater than 2,300mg per l, it seems obvious that a significant amount of sedimentary particles may adhere to these eggs. In the laboratory it is known that the hatching of striped bass eggs did not seem to be modified by suspended solids varying in concentration from 20 to 2,300mg per l. In nature there seems to be strong evidence that concentrations of sediment (of particle size 1 to 4 micrometers) of about 1,000mg per l brings about a markedly lower success of hatching of the eggs of the striped bass; there seems to be no evidence of abnormal egg development in these various suspensions. There are certain sublethal effects of a number of environmental factors that can change the survival chances of marine fish eggs. These include biochemical, physiological, morphological and behavioral effects (Rosenthal and Alderdice, 1976).

Some Conclusions on Turbidity and Fish

What can one conclude with respect to the effect of turbidity on fish biology? Certainly one conclusion that seems appropriate is that "many of our long-held beliefs as to the impacts of suspended solids and sediment on fish reproduction are based on very circumstantial evidence" (Muncy, 1979). There is a pitifully small amount of experimental data resulting from controlled laboratory experimentation upon which one can base a conclusion concerning the general overall impact of suspended solids and sediment on the reproductive success of fish. It has been maintained that a large amount of research is available in this general area of fish biology. The facts as represented by the meager quantitative literature on fish reproduction and turbidity suggest that one area in marine studies that could benefit by additional research is the relating of turbidity in all its aspects to the biological success of fish species. Specifically, as Muncy et al. (1979) have pointed out, the following seems to be an appropriate demand: "Research is needed for most species to experimentally determine the lethal and sublethal effects on all life stages of fish chronically exposed to elevated levels of suspended solids and sediment." Both chronic and acute studies in the laboratory and in well-designed quantitative field studies are important.

From a much more broad point of view it seems desirable that well-designed chronic studies be carried out in simulated estuaries or impoundments to test the manner in which suspended solids act on fish reproductive processes.

The role of runoff and of large floods moving into coastal waters should be considered in any of these studies. In this kind of quantitative work it is important to separate turbidity per se from the effects of pesticide residues and other lethal and harmful factors that ordinarily accompany suspended solids in the aquatic environment.

In connection with the problem of turbidity Mount in the report published by Swenson (1978) points out most cogently: "From the standpoint of the area affected [Lake Superior a portion of the marine environment of the United States] and the tonnage of material in the water, turbidity from silt and clay erosion is probably one of the most significant water pollution problems in the United States. Because of the multiple inputs of suspended material producing turbidity from an enormous number of man's activities, the control of turbid water is an extremely expensive and illusive matter. While there is general agreement among biologists that turbid water has adverse effects on aquatic communities, there is little information on which to prove such effects." In addition many types of turbidity render the water esthetically unpleasant to those who may view it.

From the studies made in Lake Superior it seems safe to conclude that turbidity caused by red clay has no direct effect on the survival of even the most sensitive life stages of fish in western Lake Superior. On the other hand, red clay modifies both the quality and the intensity of light in a remarkable fashion even when turbidity levels are comparatively low. It seems clear from the observational evidence that behavioral reactions of fish to turbidity and to the resulting changes in light quality and intensity have a major influence on important populations of fish. The data show that increasing turbidity causes an increase in the production of walleye in western Lake Superior "by reducing light intensity which directly enhances their feeding success and by causing rainbow smelt to become pelagic, increasing walleye food availability. In Lake Superior apparently the walleye feed virtually exclusively on smelt; walleye seem to require low light intensities and a dense pelagic population of prey in order to sustain a high rate of food consumption" (Swenson, 1978).

Sorensen et al. (1977) have summarized the effects of suspended solids in fresh water on salmonid fish (Table 3-IV).

Some Biological Responses to Decreased Turbidity

Monniot (1979) has addressed himself to adaptations in marine animals stimulated by a *decrease* of suspended particulate matter in the ocean depths. He points out that in one group of marine animals, the Ascidians, the average size decreases with increasing depth at which the specimens are found. For example in the neritic zone the size is over 1 centimeter; on the abyssal plain, less than 2.5 millimeters. Weight of the animals decreases over the same depth range by a factor of 25. Miniaturization occurs, that is, the various organs in the body decrease in size proportionally; "the further [sic] down the continental slope they are, the smaller the species of the same genus become" (Monniot and Monniot, 1977). As part of the miniaturization process specific organs become more simplified; an example of this tendency

Table 3-IV.
SUMMARY OF EFFECTS OF SUSPENDED SOLIDS ON SALMONID FISH.

Fish (Species)	Effect	Concentration of Suspended Solids	Source of Suspended Materials	Comment
Rainbow Trout (*Salmo gairdneri*)	Survived one day	80,000 ppm	Gravel washing	
	Killed in one day	160,000 ppm	Gravel washing	
	50% mortality in 3½ weeks	4,250 ppm	Gypsum	
	Killed in 20 days	1000-2500 ppm	Natural sediment	Caged in Powder River, Washington
	50% mortality in 16 weeks	200 ppm	Spruce fiber	70% mortality in 30 weeks
	1/5 mortality in 37 days	1,000 ppm	Cellulose fiber	
	No deaths in 4 weeks	553 ppm	Gypsum	
	No deaths in 9-10 weeks	200 ppm	Coal washery waste	
	20% mortality in 2-6 months	90 ppm	Kaslin and diatomaceous earth	Only slightly higher mortality than control
	No deaths in 8 months	100 ppm	Spruce fiber	
	No deaths in 8 months	50 ppm	Coal washery waste	
	No increased mortality	30 ppm	Kaslin or diatomaceous earth	
	Reduced growth	50 ppm	Wood fiber	
	Reduced growth	50 ppm	Coal washery waste	
	Fair growth	200 ppm	Coal washery waste	
	"Fin-rot" disease	270 ppm	Diatomaceous earth	
	"Fin-rot" disease	200 ppm	Wood fiber	
	"Fin-rot" disease	100 ppm	Wood fiber	Symptoms after 8 months exposure
	No "fin-rot"	50 ppm	Wood fiber	
	Reduced egg survival	(Siltation)		Eggs in gravel
	Total egg mortality in 6 days	1000-2500 ppm	Mining operations	Powder River, Oregon (Not specifically rainbow trout eggs)

Table 3-IV. (Continued)

Fish (Species)	Effect	Concentration of Suspended Solids	Source of Suspended Materials	Comment
Pacific Salmon (Oncorhynchus)	Survived 3-4 weeks	300-750 ppm (2300-6500 ppm for short periods each day)	Silt	Fingerlings
	Reduced survival of eggs Supports populations	(Silting) (Heavy loads)	Glacial silt	Eggs in gravel Spawn when silt is washed from spawning beds Yuba River, California
	Avoid during migration	(Muddy water)		
Brown Trout (Salmo trutta)	Do not dig redds	(Sediment in gravel)		Water must pass through gravel
	Reduced populations to 1/7 of clean streams	1000-6000 ppm	China-clay waste	
Cutthroat Trout (Salmo clarkii)	Abandon redds	(If silt is encountered)		
	Sought cover and stopped feeding	35 ppm		Two hours exposure
Atlantic Salmon (Salmo salar)	No effect on migration	Several thousand ppm		River Severn, British Isles
Brook Trout (Salvelinus fontinalis)	No effect on movement	(Turbidity)		

is the branchial sac which is changed anatomically. The overall biological drift is found in invertebrates other than the Ascidians (Thiel, 1975). According to Monniot (1979): "Ascidians, because their mode of food intake is based on filtering water, make it possible to hypothesize a relation between reduction in size and undernourishment." Monniot refers to von Bertalanffy's weight increase equation, $dW/dt = aW^m - kW$, where a is the anabolic constant, k is the constant of katabolism, and m is the rate of metabolism. Each factor in the equation is species specific.

The maximal value of W is associated with conditions of nutrition. In the face of chronic under-nutrition, as is characteristic of deeper waters, small size is imperative and favorable for survival and for flourishing.

Asexual reproduction in Ascidians requires the accumulation of significant energy reserves. It is apparent that asexual reproduction becomes formidable in the presence of an inferior energy budget in the animal under consideration. "The paucity of colonial forms in the abyssal zone can therefore be explained by undernourishment" (Monniot, 1979).

The distribution of larvae of Hawaiian fish was studied as related to the turbidity of the waters around the various islands of that state (Miller, 1974). Secchi disk values were not used other than to divide roughly the results: Miller holds that the precision of the Secchi disk transparency measurements is very low, about ± 10 to 20 percent. Fish were put into two classes; those in turbid water and those in clear water. The former were from water in which the disk depth was less than the water depth. The number of fish larvae per $1000m^3$ water was about 75 percent less in turbid water than in clear water. The number of species represented was lower by about 55 percent in turbid water. "Turbidity, whether natural or artificial, was negatively correlated with larval fish abundance" (Miller, 1974).

The turbidity effect seems to be real (not an artifact). Why then the correlation found? The easiest answer is that in water of greater turbidity fish larvae are unable to see; as a consequence they cannot move counter to the currents; thus they do not accumulate. Preliminary observations suggest that fish larvae cannot accumulate in waters with a mean current speed of more than 25 centimeters per second (Table 3-V).

Table 3-V

TABLE SHOWING THE MEDIAN VALUES FOR DENSITY, NUMBER OF SPECIES, AND DIVERSITY FOR LARVAL FISH IN CLEAR AND TURBID WATER RESPECTIVELY. S, SUMMER OBSERVATIONS; W, WINTER OBSERVATIONS. AFTER MILLER (1974).

Observation	Clear water S	Clear water W	Turbid water S	Turbid water W
No. of larvae 1000/m^3	180	115	60	20
No. of species X 10	400	300	180	110
Diversity X 10^2	390	390	325	270

Pollution Considerations

Eisma (1981) has discussed suspended particulate matter as a carrier medium for pollutants in estuaries and in the open ocean. He points out that particles suspended in the ocean range in size from 0.02m up to several millimeters in diameter. Despite the fact that less than 5 or 10 percent of the suspended particulate matter that is supplied by continental land actually reaches the deep ocean, what does reach the deep ocean is postulated to be relatively important for the transport of pollutants from the land to the ocean. This view is based on the small size and the relative high content of absorbed material characteristic of these deep ocean suspended particulate materials.

Anderson (1980) has discussed the influence of tidal flow over a tidal flat on suspended particulate matter in such a region.

Coral Communities

Loya (1976) studied the community structure of corals off the west coast of Puerto Rico. The results of the study suggest that turbidity of the water as well as sedimentation are probably the major factors that determine the distribution of corals in the different reef zones. Specifically, where the coral diversity was great the average water turbidity was rather low, 1.5 Formazin turbidity units; where coral diversity was low, the average turbidity of the water was significantly higher, 5.5 FTU.

Chapter 4

METHODOLOGY

Continuous Monitoring

ARCHER (1980) has reviewed sensing elements that can be used for the continuous monitoring of turbidity in "water."

Gordon and Smith (1972) designed and installed a continuous reading 45° angle light scattering meter in the Florida Current. Results showed a high concentration of suspended particulate matter near the bottom. "Near the axis of the current, an anomalously high scattering coefficient, which may be an indication of upwelling, is observed as well as a scattering maximum near the depth of largest vertical shear." The scattering coefficient in the uppermost water layer was about 0.05 per meter.

Shiroto (1979) has concluded that turbidity values, measured at 400nm after 1 minute standing in an absorption cell, reflect very well the water masses in the Ariake Sea as revealed by temperature-salinity (T-S) diagrams. He suggests that turbidity change may be used to indicate distinct water masses in an estuary.

Permanent turbidity standards have been prepared by suspending finely divided titanium dioxide in aryl sulfonamide-formaldehyde or methylstyrene resins (Roessler and Brewer, 1967). They seem to have a useful life for at least twelve years.

Remote Sensing and Turbidity

Gierloff-Emden (1977) has published a manual for the interpretation of remote sensing done in coastal and off-shore environments.

The optical characteristics of water in the ocean determine how deep light will penetrate into the sea and the spectral composition of the light at different depths. Indirectly then these characteristics have an effect on the temperature of the ocean, on the color of the ocean and on the primary production of phytoplankton. Consequently it is evident that all marine life is affected by the optical properties of seawater. These properties can be truly called crucial; they include reflection, refraction, dispersion, and absorption. "The top ten meters of seawater in the open ocean absorb between 33 and 80 percent of the blue-light depending on the clarity of the water" (Gierloff-Emden, 1977).

The transparency of ocean water has been for many years measured by the Secchi disk method. This procedure measures the penetration of light from above down into the sea. This aspect of transparency is not the same as the total light penetration which is perceived by a diver or by a submarine at some known water depth. The total penetration of light into the ocean is greater than the Secchi disk transparency value; the difference is 10 to 100 times.

"Illumination at increasing depths follows an approximately exponential law and depends on many parameters" (Gierloff-Emden, 1977).

The waters of the open ocean and in the near coastal regions consist of a mixture of various materials in a solution, materials in colloidal suspension, and the suspension of particulate matters as well as turbidity caused by inorganic and organic substances. There is also a so-called "yellow substance" that gives a basic color to these near-shore waters. Turbidity is clearly a complex, optical, physical characteristic of water.

Interpretation of Remote Sensing Data

In the interpretation manual Gierloff-Emden (1977) uses several striking examples of interpretation of optical properties of seawater from remote sensing, i.e. satellite observation. It is obvious that color photography from satellites requires different standards of interpretation than the usual photographs taken by individuals standing near a body of water. Nevertheless the information available from such satellite observations is impressive. Masses of water can be distinguished in lagoons and coastal areas characterized by extremely transparent water. Bottom configurations can be studied with great profit from remote sensing posts.

It is suggested that individuals who are concerned with matters of water turbidity, transparency, and suspended particulate matter will profit by a study of the Gierloff-Emden manual. An excellent list of references is given to support the information made available in this book. The manual demonstrates quite clearly that remote sensing particularly of coastal and offshore waters is an important research and information gathering tool which permits the detailed repeated observation of remote bodies of water frequently enough so that new and dramatic information can be gathered. In most instances it is impossible or certainly not economically feasible to make in-depth repeated studies of optical properties of even a single bay or river mouth using the historically accepted procedures of oceanography. With increasing sophistication of remote sensing equipment it now becomes possible to make such observations. Consequently a whole new world of information about the optical properties of coastal and estuarine waters is now uncovered. New approaches and new precision to understanding the movement of water masses throughout the coastal and offshore regions of the world are now becoming readily available to oceanographers in need of this information. In my estimation we are seeing only the very beginning of a dramatic growth of remote sensing in the study of the ocean. It does not seem outlandish to suggest that a wide variety of chemical and physical tests will be run repeatedly and with great precision using remote satellite sensing and other similar devices.

Miller et al. (1977) analyzed airborne spectral measurements along three separate lines of flight over Georgian Bay. The results were used to estimate the chlorophyll concentration in the waters that were flown over.

Light Penetration of Water

Years ago Harvey (1960) clearly pointed out the importance of light penetration into the ocean in the determination of the growth of phytoplankton. He pointed out moreover that at local noon direct sunlight that has a wavelength of 360-760 m has a different spectral composition than does the light that is coming directly from the sky. Both direct sunlight and light from the sky change composition as the sun moves toward the horizon. "Thus the light falling on the sea is continually changing in spectral composition." These facts can be illustrated by the following data. Light of a violet blue color, wavelength 380-490 m makes up about 27 percent of the mean noon sunlight; it makes up about 50 percent of the blue sky light. Other wavelengths of light give the following percentage compositions for mean noon sunlight and blue sky light respectively. Green, 490-560 m, 24 percent of mean noon sunlight; 24 percent of blue sky light. Yellow and orange, 560-620 m, 20 percent of mean noon sunlight; 12 percent of blue sky light. Finally red light of wavelength 620 - 720 m makes up about 29 percent of mean noon sunlight and 13 percent of blue sky light.

The amount of light that is reflected on the surface of the sea varies from about 4 percent during calm weather on a sunny day to as much as 25 percent when the sun is shining and there is a moderate sea. During an overcast day approximately 8 percent may be reflected. Blue light in general is scattered more than is red light but it is not as readily absorbed in ocean water as is red light.

Harvey's (1960) discussion of turbidity and water transparency made two decades ago is still worth reading.

Remote Sensing – Photographic Imagery

Casual or *ad hoc* photographs, some in color, taken from orbiting vehicles have given a hint that differentiation of water masses on the basis of turbidity or variations in chlorophyll content may be practical. The many variables accompanying the *ad hoc* picture taking virtually shut off the possibility of orderly investigations based on that methodology (Apel, 1978). Moreover, ordinary color photography is not readily amenable to multispectral analyses.

"Nevertheless, considerable qualitative and semi-quantitative information on coastal flows, turbidity plumes, plankton concentrations and roughness variations may be derived from a relatively simple analysis of the colour imagery" (Apel, 1978; Gierloff-Emden, 1976).

The color photographs in the Gierloff-Emden (1976) manual reveal some striking information. Water discolorations of coastal and estuarine waters are revealed on a broad scale. Surface optical characteristics literally shine forth. Certainly an orbiting vehicle offers the oceanographic investigator an overall better vantage point than does the deck of a ship.

What does the present state of the art in orbiting photographic imagery offer to the investigator of oceanic turbidity? High resolution color photography under carefully selected conditions can reveal the following: areas of phytoplanktonic blooms; red tide areas; turbidity plumes originating from rivers, beaches, inlets, glaciers, and dredging operations; sewage sludge dumping.

Quantitative data, using this method, are restricted to geometric or geographical spread of the item observed. Data on concentration, depth, height and similar quantities must wait upon future developments in instrumentation.

There is a large collection of photographs available from previous orbited and high altitude photographic missions. Many of these will support research efforts of aquatic scientists. For a catalogue of oceanographic satellite data refer to Brunn Memorial Lectures (1978).

Specific turbidity relationships of color imagery* may be summarized as follows:

Turbidity

In the context of this discussion, "turbidity" will imply increased scattering of light from nonliving particulate matter in the water column, chiefly sediments and *gelbstoffe*. Turbidity plumes are readily visible in Landsat images at the green end of the spectrum.

There are many sources for such scatterers. River outflows, bottom material placed in suspension by wave action, estuarine waters flowing out through inlets, glacier "flour" and dredging operations have been documented as sources of increased oceanic turbidity. There are preliminary indications that the near-surface concentration of suspended

---

*Apel (1978) expresses the opinion that: "In the light of the superiority of other sensors, there does not appear to be any significant oceanographic need for camera imagery in the future."

sediments may be inferred from measurements of the radiances in two or three Landsat channels, provided the general type of sediment is known, e.g. its optical index of refraction and its average size. Calculations applying Mie scattering theory to the problem of light welling up from the sea indicate that a larger number of selected narrow spectral hands will yield good estimates of the concentration; again the Nimbus-G Coastal Zone Colour Scanner has been designed with this application in mind (Apel, 1978).

Attenuation Vis à vis Turbidity

Russian oceanographers contend that attenuation measurements can lead to valid estimates of amount of suspended matter in a region of the sea, including plankton (Nelepo, 1978; Kondratev and others, 1972). In order to obtain accurate, interpretable results from spacecraft it is essential to make simultaneous observations (photographs) of the ocean surface in three or more frequency bands (Sagdeen, 1977).

Mead Lake Remote Sensing

A data acquisition and analysis program has been undertaken to demonstrate the feasibility of remote multispectral techniques for monitoring suspended sediment concentrations in natural water bodies. Two hundred surface albedo measurements (400 to 1,000 nanometers) were made at Lake Mead with coincident water sampling for laboratory analysis. Water volume spectral reflectance was calculated from the recorded surface albedo, and volume reflectance-suspended sediment relationships were investigated. Statistical analysis has shown that quantitative estimates of nonfilterable residue and nephelometric turbidity can be made from volume spectral reflectance data with sufficient accuracy to make the multispectral technique feasible for sediment monitoring (Holyer, 1980).

Monitoring

Turbidity measurements by the use of monitoring sensors cover typically, in estuarine waters, a concentration range of 0-100 Formazin Turbidity Units (FTU). The sensor in question in many instances is spectrophotometric. The overall range of the sensor for turbidity is 0-400 (FTU). Typically light at 510nm is used.

Total suspended solids in estuaries are typically in the range of 10-150 mg per l with an accuracy of about 1.5mg per l (Archer, 1980).

For total organic carbon automatic monitoring devices based on the demonstrated good correlation between ultraviolet absorption at 254nm and total organic carbon are ideal.

NASA Plans

The National Aeronautics and Space Administration has recently (March 2, 1981) let three design study contracts for developing a National Oceanic Satellite System to Lockheed Missiles and Space Company, Radio Corporation of America, and Rockwell International respectively. The $750,000 study is the first step in development of a limited pilot system for remote sensing of oceanic conditions from space. A color scanner will be included in the monitor and should reveal optical characteristics of the ocean on a wide scale.

Some Developments in Methods

Shepherd (1978) has described a rapid method for the automatic graphical representation of depth-turbidity-salinity.

New advances in methodology have been discussed by several investigators. Submarine radiometry and environmental monitoring were evaluated by Vandenberghe (1978). Secchi disk values in the northeastern Mediterranean Sea have been reconsidered (Collins and Banner, 1979). Modeling of turbidity flow has been attempted (Pantin, 1979). A rapid method for correlating depth, turbidity, and salinity in order to obtain a graphical output has been suggested (Shepherd, 1978). Probst (1976) has published a descriptive model for particulate cycling in the Baltic Sea. A new transparency meter for studying underwater light penetration was described by Vishnudatta and Murty (1978). Numerical simulation of turbidity has been applied to the Gironde Estuary (Penhoat and Salomon, 1979).

Ghovaniou et al. (1973) evaluated the effectiveness of an airborne laser-detection system, using blue-green laser light. Turbid waters were recorded to have an attenuation coefficient from 0.07/m to 2.0/m.

Instrumentation

Hach Chemical Company has published discussions of turbidity (Hach, 1979; Eichner and Hach, 1971) based on Formazin Turbidity Units (FTU). These units now have wide adoption by industry and by the American Public Health Association. Suspensions of Formazin may be made with great exactness to be used for calibrating turbidimeters. FTU now replaces the Jackson Turbidity Unit (JTU). The units are interchangeable.

Eichner and Hach (1971) reported the value of 0.022 FTU for the turbidity of ultra-pure water using a quartz iodine light source.

Hach (1979) reviews the history of turbidity measurements. Nephelometry is the method of choice for measuring turbidity with the Formazin suspension as the primary reference. Several cautions are important in cross comparing reported turbidity values. It is dangerous to relate cursorily turbidity measurements to the value mg/liter of suspended material. Too many samples of water do not show a linear relation between mg/l suspended material and measured turbidity values. If samples of water are diluted, some of them result in the suspended particulate matter breaking up into a large number of smaller diameter particles. Moreover, variations of 5X can result if different turbidimeters, all standardized against the same Formazin suspension, are used to measure the turbidity of a turbid suspension other than Formazin.

The United States Environmental Protection Agency (1979) has approved the nephelometric method for measuring turbidity: "This method is applicable to drinking, surface, and saline waters in the range of turbidity from 0 to 40 nephelometric turbidity units (NTU). Higher values may be obtained with dilution of the sample." NTU = FTU = JTU.

The Hach Turbidimeter, Model 2100 and 2100A, or equal from other sources is accepted by EPA for the nephelometric test.

The scanning electron microscope has been shown to be of special usefulness in characterizing suspended particulate matter in oceanic and in estuarine waters (Pierce and Siegel, 1979).

United States Federal Government Recommended Method

The National Handbook of Recommended Methods for Water-Data Acquisition recognizes that the measurement of turbidity has been a common practice in water analysis for many years. Moreover, it admits that "attempts to quantify turbidity have led to a proliferation of methods and units." In the handbook it is, consequently, held that the term "turbidity" is widely taken as a nontechnical qualitative descriptive word for the optical characteristics of water. By agreement among several federal analytical laboratories it was decided to discard all attempts to measure turbidity. The replacement consists of measuring, using white light, transmission and 90 degree scattering.

The demonstrated accuracy of the system used for the scattering measurements must be ± 5 percent of full scale over a range of temperature from -2° to 40°C. The time constant of the system must not exceed 60 seconds. Calibration must hold, with clean surfaces in the system, for at least four weeks. For the transmittance system of measurement the same specifications must obtain.

To make these measurements, an appropriate sample of water is collected in a glass or plastic container. No preliminary treatment or preservative is required. Samples should be stored at 4°C and analyzed as soon as possible. The method of analysis is nephelometric. The usual range will be 0 to 40 Formazin Turbidity Units; higher values are obtainable if the sample is diluted. Significant figures are as follows:

1-10 FTU	nearest 0.1 FTU
10-40 FTU	nearest 1.0 FTU
40-100 FTU	nearest 5.0 FTU
100-400 FTU	nearest 10.0 FTU
400-1000 FTU	nearest 50.0 FTU
Greater than 1000 FTU	nearest 100.0 FTU

Equating Laboratory with Field Observations

There are some problems with equating laboratory observations of turbid waters with direct observations in the field. For example studies on polarization of turbid suspensions in the laboratory indicate that the angular distribution of the extent of polarization is *not* described by the following equation (Timofeeva, 1961):

$$p = \frac{1 - \cos^2 \theta}{1 + \cos^2 \theta} \frac{\sin^2 \theta}{1 + \cos^2 \theta}$$

where p is the degree of polarization. At angle $\theta = \pi/2$ complete polarization obtains, angle θ is direction of radiant intensity.

In the sea, direct observations demonstrate that greatest degree of polarization obtains at θ = near 90°, regardless of wavelength of light considered; the equation for p above clearly applies in direct observations at sea (Jerlov, 1968).

Fresh Water Turbidity Studies

For various reasons the study of turbidity in fresh water systems is muddled a bit by definitions. Turbidity has been defined elsewhere in this book. In the marine environment the application of the term is reasonably straightforward.

In the freshwater environment other terms, such as suspended matter, suspended solids, suspended sediment, bed sediment, are used as well as turbidity. The cross-reference conversion factors are not always clear (Farnworth and others, 1979).

Measurements of turbidity in the Lot River, France, give data of use in comparing suspended matter to turbidity (Decamps and others, 1979). Total suspended matter in mg/l is made up of 4.4 ± 0.8 mineral matter and 5.5 ± 0.7 organic matter. The comparable turbidity in Formazin Turbidity Units was 4.4 ± 0.4 FTU.

Decamps et al. (1979, p. 282) have published a useful diagram which illustrates the dynamics of suspended matter in the Lot River.

Webster et al. (1979) have published a tested model for predicting the effects of impoundments on the movement of particulate organic matter in a river system.

Chapter 5

DYNAMICS OF SUSPENDED PARTICLES

The Dynamic Role of Suspended Particulates

IT is important to realize that particulate material in the ocean is not a mass of inactive dead static substance — quite to the contrary. Suspended particulate matter has an important dynamic role to play in the ocean.

Turekian (1977) has reviewed the fate of particulate elements in the sea. He concludes that "The major sequestering of trace metals in the oceans is mediated through particles." Indeed, studies of ^{210}Pb in the ocean surface reveal that particulate material scavenges finally many trace metals more importantly than do individual living organisms. The freedom of the open oceans from metals is readily explainable by the unrelenting action of suspended, or resuspended, particles that scrub down, and reduce, trace metals from the water column to burial in the bottom complex.

The role of resuspension of particulate matter is only now being fully appreciated as a dynamic force in the flux of many materials from the bottom layers of the ocean into the upper layers of the sediment. It is apparent, however, that one must now look upon suspended particulate matter as an important factor in the chemical constancy of composition in the ocean and the composition of the sediments on the bottom of the ocean. The idea of suspended particulates being merely suspended inorganic material having no apparent function, no longer is an acceptable conclusion. There is much more research to be done to ascertain the details of how these suspended particulates behave and marine scientists must recognize that things are happening to suspended particles during their entire existence until burial in the bottom sediments. Moreover, things are happening in the individual particles themselves. Additionally, these particles do various things to dissolved material substances. It is obvious then that the suspended particulate system in the ocean is a complex one that deserves intensive and extensive study over the next decades to clarify and expand the unique and fundamentally important function that suspended particulate matter plays in the world ocean.

Some Characteristics of Oceanic Particulates

Gardner (1977) has made a detailed experimental study of the various characteristics of particulates found in the ocean. He used sediment traps that were constructed to reveal quantitative information about the movement of particulates in specific zones of the ocean. Using the technique of sediment traps, it was found that somewhere from 80 to 90 percent of the particulate matter collected had particle sizes that measured less than 63 micrometers. The average diameter of particulate matter that was sampled in the nepheloid layer was about 20 micrometers; that from above the nepheloid layer had an average of about 11 micrometers in diameter.

"Less than 3 percent of the organic carbon produced in the photic zone at the trap sites was collected as primary flux 500m above the sea floor. The primary flux measured at two sites was enough to supply 75 percent on the upper Rise and 160 percent [sic] on the mid Rise of the organic carbon needed for respiration and for burial in the accumulating sediments" (Gardner, 1977).

Large particles apparently carry carbonate and organic matter preferentially; these particles fall rapidly toward the bottom of a column of ocean water.

The nepheloid layer is kept up essentially by the resuspension of sediment. Without this process the nepheloid layer would persist at best for a few months. The time of residence for particulate material in this nepheloid layer is numbered in days. In the layer a few meters thick just above the ocean bottom, the residence time for the total nepheloid layer is calculated to be weeks or even months.

Nepheloid Layer

The nepheloid layer of the ocean is a dynamic part; deposition of material with resuspension of this material takes place constantly. It is incorrect to look upon the nepheloid layer as a static body of water containing a wealth of particulate matter which is held indefinitely in suspension. There is a rapid recycling of particles; thus it is that particles are resuspended over and over again; eventually they are buried completely and are incorporated into the sediment at a level where they are not readily reexposed. The process of resuspension makes it possible for particulate matter to be carried longer distances than would be likely if particles made a single trip through the column of water. By the same token the resuspension of particulate matter exposes particles of carbonate and of silica, respectively, to the corrosive action of seawater, which may contain less than a saturation level of these specific ions. This set of circumstances could be a critical control method for dissolution rates of such particulate matter.

Particulates as Scavengers

In connection with this matter of the scavenging action of particulate materials Minagawa and Tsunogai (1980) suggest "that suspended particles are somehow closely related to the removal of heavy metals from seawater, in spite of the negligibly small settling flux of suspended matter." Their work was based on equilibrium studies between ^{234}Th and ^{238}U and particulates in Funka Bay, Japan.

In surface Pacific Ocean water the amount of ^{10}Be is about 740 atoms per cubic centimeter, a value significantly lower than the amount of Be in precipitation. The recent study on scavenging (Raisbeck and others, 1979) concludes that the scavenging of Be by particulate matter is an important oceanic surface water process. Nuclear accelerator spectrometry was the method of measurement; it proved to be feasible for detailed vertical and horizontal distribution studies of ^{10}Be in the ocean.

Particulate Organic Matter

Suspended particulate organic matter plays a crucial role in the transport vertically of matter in the ocean (Saino and Hattori, 1980). Suspended particulate organic matter is produced by the phytoplankton of the euphotic zone in the open ocean. This particulate matter is broken down in the aphotic layer while it sinks downward to the floor of the ocean. "Isotopic abundance of biophilic elements such as C and N in POM is altered by isotopic fractionations associated with biochemical reactions." For this reason C-13 or N-15 can serve as useful sources of information concerning how suspended particulate organic matter behaves in the ocean. It seems that there are two principal routes for the transport vertically of particulate matter in the ocean. First of all, transportation is carried out by slowly sinking fine suspended particulate matter. This material is readily collected on filters by passing small amounts of seawater through them. Secondly, transportation is carried out by the rapid sinking of larger particles; an example of these would be fecal material. The usual collecting methods are not adequate. However, this material can be collected fairly satisfactorily by filtering large amounts of seawater or by using so-called sediment traps. The measurement of various components of suspended particulate organic material may provide a message for evaluating the importance of various processes in the vertical transportation of particulate matter. Radioactive isotopes of nitrogen and of carbon are important in this regard.

DiToro (1978) has applied the important equations of optics to a large amount of data collected in the San Francisco Bay Estuary. Using this approach, which is purely mathematical, it was possible to estimate the optical properties of particles as functions of the concentration of nonvolatile suspended solids, algal chlorophyll, and suspended detritus. The mathematical methods seem to be appropriate and effective for developing relationships between the optical properties of a given body of water and the characteristics and concentrations of particulate matter.

Amino Acids in Suspended Solids

Particulate matter from different portions of the Pacific Ocean has been shown to contain some fifteen amino acids (Siezen and Mague, 1978). Particulate matter from various regions and depths in the Pacific Ocean was hydrolyzed and the amount of different amino acids contained was measured by gas-liquid chromatography. In the euphotic zone the relative composition for amino acids was quite alike in all samples analyzed. The total particulate amino acid concentration had a range of 370-2260 nmoles/liter in water from coastal zones; 90-260 nmoles/liter in the open ocean. With increasing depths, the concentration of particulate amino acids quickly decreased; below a depth of 200 meters the concentration of particulate amino acid leveled off at 10-40nmoles/liter. In the deep waters of the equatorial region and in the deep ocean off the coast of California the most abundant individual amino acids in particulate matter were glycine, serine, glutamic acid, and aspartic acid. In many samples all the particulate organic nitrogen found could be accounted for by the nitrogen content of the particulate amino acids; particulate amino acid carbon "contributed up to 50 percent of the total particulate organic carbon in euphotic waters and down to 20 percent in deep waters." In near surface water it was calculated that 11-32 micrograms per liter of protein was equivalent to the total particulate amino acid content at open ocean collecting points. At coastal stations about 270 micrograms per liter of protein was calculated to be equivalent to the total particulate amino acid content in the near surface waters.

The authors maintain that their work "indicates that an indirect measurement is probably valid for estimating the protein content of POM [particulate organic matter]. However, a knowledge of the total amino acid composition provides greater insight into the nature of POM and the transforming processes which modify it" (Siezen and Mague, 1978).

How Particulates Adsorb Metals

The adsorption of trace metals by amorphous hydrous ferric oxide in seawater is an interesting phenomenon with implications for water pollution, aquatic chemistry, and the chemical dynamics of estuaries. Morel and Swallow (1980) used a simple model to study such adsorption. Their research reveals that this complex system acts in a noteworthy simple manner. Experimental and observational results gathered under different circumstances of total metal concentration and total oxide concentration are reducible to a reaction constant K, which is a function solely of hydrogen ion concentration. The great capacity of the oxide for trace metals denies as inadequate the view of a surface reaction to explain the uptake of metals. The authors present a physical image of the oxide as a swollen hydrous gel permeable to hydrated ions.

Chapter 6

ESTUARIES

Definition

AN estuary may be defined as follows: "An estuary is a semi-enclosed body of water — a bay, a lagoon, or the mouth of a river — in which fresh and salt waters meet and mix. Along the thousands of miles of the nation's coasts, there are many estuaries, some large and some small. Estuaries are critical to the food chain of many species of life" (Council on Environmental Quality, 1981).

Water Characteristics

The Council on Environmental Quality uses thresholds of various water characteristics to ascertain water quality of estuaries. Turbidity and chlorophyll *a* values are considered not applicable for that purpose (Table 6-I).

Yarra Estuary

There is a strong correlation between trace metals and iron in the sediments of the Yarra Estuary, Australia (Smith and Milne, 1979). This fact suggested that the activity of *Fe* in suspended particulate matter should uncover data of value in furthering understanding of how trace metals are removed from those waters.

Most of the suspended particulate matter (expressed as gram/meter^3) varied with salinity in the estuary. A vertical and a horizontal sampling respectively gave strikingly different relationships of particulate matter concentration to salinity.

Apparently in the vertical section suspended particles are at first removed slowly as salinity increases. On the contrary in the

Table 6-I.

THRESHOLDS USED IN THE CEQ ANALYSIS OF NATIONAL ESTUARY WATER QUALITY (COUNCIL ON ENVIRONMENTAL QUALITY, 1981).

Conductivity	COND	NA
Turbidity	TURB	NA
Salinity	SAL	NA
Chlorides	Cl	NA
Biochemical oxygen demand	BOD	5 mg/l[a]
Fecal coliform bacteria	FC	14 cells/100ml[b]
Chlorophyll a	Chl[a]	NA
Nitrates	NO$_3$	10 mg/l[c]
Total kjeldahl nitrogen	TKN	40 mg/l[c]
Phosphates	PO$_4$	1 mg/l[a]
Total phosphorus	TP	1 mg/l[a]
Total cadmium	Cd	1 g/l[d]
Total lead	Pb	10 g/l[e]
Total mercury	Hg	0.1 g/l[f]
Total arsenic	As	29 g/l[d]
Total chromium	Cr	25 g/l[d]
Total copper	Cu	5 g/l[d]
Total zinc	Zn	25 g/l[e]

l = liter; ml = milliliter; mg = milligram; g = microgram.

NA = Not applicable.

[a]Value chosen by CEQ because relatively high concentration of this indicator is not critical to estuaries.

[b]EPA criteria level for "shellfish harvesting waters."

[c]Value used in the absence of an EPA *Redbook* or proposed criteria level and chosen by CEQ after discussions with EPA about a reasonable value to use.

[d]EPA proposed criteria level for preservation of aquatic life in estuarine waters.

[e]Value used in the absence of an EPA *Redbook* or proposed criteria level and chosen by CEQ after discussions with EPA about a possible final criteria level value.

[f]EPA criteria level for preservation of aquatic life in marine waters.

The EPA *Redbook* is: U.S. Environmental Protection Agency, *Quality Criteria for Water* (Washington, D.C.: U.S. Government Printing Office, 1976).

surface waters suspended particulate matter is removed rapidly at first as the salinity increases. The two curves approach one another as salinity increases:

	Particulates, g/m³	
Salinity, 0/00	Surface Waters	Vertical Section
1	39	41
10	20	38
15	15	35
20	10	32
28	4	26
32	-	3

The content of *Fe* extractable in suspended particulates if plotted against salinity gave a single fairly smooth, symmetrical curve no matter the source of the sample:

Salinity, 0/00	Fe, mg/gm
1	10
5	15
10	20
15	22
20	21
25	17
28	10

Smith and Milne (1979) maintain that the strong correlation of extractable *Fe* concentration with lead and zinc concentrations in sediments plus the capacity of *Fe* "to form particles or coatings of hydrated ferric oxides in estuarine conditions, indicate that the behaviour of *Fe* is a major factor in controlling trace metal concentrations in sediments." Copper is not associated with iron in this respect. According to these investigators: "The behaviour of *Cu* indicated that it was not only controlled by the hydrated iron oxides, but that some may be organically bound." There was an average of about 10 milligrams of iron per gram of particulate in the essentially fresh waters of the upper estuary.

Narragansett Bay

Davis (1973) published the results of a turbidity survey made in Narragansett Bay, Rhode Island. During the summer variations are great. One standard deviation at any specific location where samples are taken makes up 7-23 percent of the mean value. The variation seems to depend on the state of the tidal current, the depth of the water at the collecting site, or the weather. The magnitude of the variation in turbidity is about four times over a north-to-south range of some 31 kilometers in the estuary. The cleanest water is at the south mouth of the Bay. There is a good correlation between turbidity values and the autumn values of suspended material concentration.

Gironde River

The estuary of the Gironde River has a zone of maximal turbidity. Radio-tracer studies have indicated the following:

1. The turbidity maximum which has been observed is not constant in volume during a semi-diurnal tidal cycle. It pulsates with the increase and the decrease of tidal currents.
2. There is an increase in the concentration of suspended sediment; and this increase is directly proportional to the velocity of currents near the bottom; it is also a function of the degree of compaction of the bottom sediment (Allen et al., 1977). A general model of the Gironde Estuary is presented. One feature is the large-scale lateral transportation of suspended sediment that is observed in the lower estuary.

Mathematical Modelling

Di Toro has made an attempt to analyze mathematically the radiative transfer in estuarine waters that are turbid; he indicates that it is a mathematical study of the "radiative transfer equation ... which is applicable for the case of angle-independent, exponentially decreasing irradiance." If reflectance ratios that are less than 10 percent are considered first, the extinction coefficient for down-welling irradiance is attributable to absorption by particles in suspension in the water column plus that fraction of light that is scattered by particles but is not scattered directly forward (Di Toro, 1978).

Equations for extinction coefficient and for reflectance ratio are compared with various approximations. From "single" scattering approximation, solar angle dependence is derived. The author applies the resulting equations to numerical data obtained in observations on the San Francisco Bay Estuary. In this way it was possible to estimate the optical properties of the particles in that body of water as functions of the amount of nonvolatile suspended solids, the algal chlorophyll, and suspended detritus.

The predicted value of the regression equations was verified by using independent data. The author contends that his methods are apparently appropriate to develop relationships between the optical properties of water and particle characteristics and particle concentrations.

Fort Pierce Inlet, Florida

The relationships among chemical, physical, and optical characteristics of a column of water which is organically rich, as well as being highly turbid, was subjected to multifactor regression analysis; the water was obtained from Florida's Fort Pierce Inlet. Measurements were made of water transparency at 445nm, 542nm and 630nm respectively. The data indicate that scattering of light by suspended particles is the principle optical mechanism that controls down-welling irradiance at all the wavelengths tested (Thompson et al., 1979). "Scattering by suspended particulate material was found to be the primary optical mechanism controlling downwelling irradiance at all three wavelengths."

If the large particles had diameters greater than 3-5 micra their scattering efficiencies were constantly 2; variations depended on the wavelengths of light. "Selective absorption had a definite effect on the transmission of radiant energy in the 445-nm range." A strong correlation was uncovered between extinction measured at 445nm and the area of the cross-section of the suspended particulate matter. This finding shows that particulate matter and not materials in solution is the major constituent in the water column that absorbs on a selective basis short wavelength radiant energy in the Fort Pierce Inlet. The light spectrum distribution of downwelling spectral irradiance shifted in a remarkable fashion during several months. These changes apparently are related to seasonal and other factors in the concentrations of compounds that are selectively absorbing in the water column under consideration. Odd events such as dredging or marine construction might also cause shifts of selectively absorbing compounds.

Maxima and Minima

Festa and Hansen (1978) devised a steady state numerical model in an attempt to explain the existence and location of turbidity maxima in estuaries that are partially mixed. They found that estuarine dynamics is chiefly responsible for the location and the magnitude of the turbidity maxima. These maxima are not related in a simple fashion to the amount of sediment introduced into estuaries by way of rivers that feed into those bodies of water. Rather a number of factors such as settling velocity (particle size) of the suspended sediment, the amount of sediment and its input from the oceanic and the river ends of the estuary respectively as well as the strength of circulation within the estuary all combine to establish the site and the magnitude of the turbidity maxima.

Weser Estuary, Turbidity Maximum

In the Weser Estuary as an example, the turbidity maximum is determined by estuarine and non-estuarine factors. The maximum exists at the tip of the salt wedge. Wellershaus (1981) concludes with respect to the Weser Estuary: "Non-tidal estuarine circulation, resulting in collection, amassment and sedimentation of particles with low settling velocities . . . seems to be the main cause for the mud character of the sediment." Ebb and flow of the tide cause alternate sedimentation and resuspension of particulate matter; the latter accounts for the discernible turbidity maximum. Finally it is concluded that:

"The behaviour of suspended particles (seston) is markedly different during neap and spring tides. Additional complicatory factors are the occurrence of a range of particles with different settling velocities, the removal of sediment by extraordinary forces like storms and freshets, and technical measures like dredging or water buildings" (Wellershaus, 1981).

Suspended Material in Estuaries

The concentration by weight of the suspended inorganic material in estuaries has been reported on numerous occasions. Concentrations ranging between 300 and 2000 milligrams per liter were measured in the Thames Estuary. Concentrations ranging from below 50 milligrams per liter in the seaward limits of the Thames Estuary to 50,000 in the mud reaches have been recorded. The higher value was thought to be due to the formation of fluid mud at the sampling site, and that a value of the order of 3000 milligrams per liter was more representative of the material in suspension.

In the upper reaches of the Mersey it was found that concentrations of 2000 milligrams per liter were common with occasional values in excess of this figure, while in the Solent Estuaries concentrations varying from 22 to 200 milligrams per liter had been measured.

The mean diameter of a typical Thames mud was given by Inglis and Allen (1957) to be 35 micrometers with about 5 percent of the sample being less than five micrometers (m) diameter. Some samples however contained 50 percent of material finer than two micrometers diameter. Samples obtained from the Mersey show that 40 percent of the material is of clay size. The sizes referred to above apply to the dispersed material but clay is usually present as flocs.

Tests made to determine the settling velocity of silt in suspension in the Thames showed that under conditions of turbulent flow salinity had little effect on the settling velocity of Thames silt and that during spring tides the settling velocity increased linearly with concentration but on neap tides it varied approximately as the square of the concentration.

The concentration of material in suspension varies during the tidal cycle and also on a seasonal basis. The development of instruments to monitor sediments in suspension has facilitated the compilation of long-term records. Analysis of records obtained in this way for the river Humber shows that there is a linear relationship between the concentration of sediment and the tidal range and that it appears to vary inversely with temperature.

Sholkovitz (1979) has concluded that the suspended matter in the Tay Estuary (Great Britain) derives from the introduction, periodically, of heavy loads of suspended particulate matter and of large volume river water discharge.

The ecology of an estuary is dependent upon the turbidity of the water and will be affected by any change in the sedimentary regime (Nihoul, 1978).

Monitoring

Turbidity measurements by the use of monitoring sensors cover typically, in estuarine waters, a concentration range of 0-100 Formazin Turbidity Units (FTU). The sensor in question in many instances is spectrophotometric. The overall range of the sensor for turbidity is 0-400 (FTD). Typically light at 510nm is used.

Total suspended solids in estuaries are typically in the range of 10-150 milligrams per liter with an accuracy of about 1.5 milligrams per liter (Archer, 1980).

Pollution Factors

Helliwell and Bossanyi (1975) have edited a book that contains a good discussion of turbidity and the role of suspended material in coastal and estuarine waters. Coastal waters generally have a small amount of suspended material compared to the amount usually found in an estuary. According to one report the inorganic suspended material in the English Channel varies over the year from 0.45-2.7 milligrams per liter. There is suspended in the waters of the English Channel, of course, organic components. Organic detritus may make up as much as 90 percent of the suspended organic matter.

The particle size of the suspended material in the Thames has been reported to be on the average 35 micrometers; 5 percent of the total particles counted were less than 5 micrometers in diameter. These size distributions apparently vary rather widely depending on the place of sampling and the technique of sampling.

"Tests made to determine the settling velocity of silt in suspension in the Thames showed that under conditions of turbulent flow salinity had little effect on the settling velocity of Thames silt and that during spring tides the settling velocity increased linearly with concentration but on neap tides it varied approximately as the square of the concentration."

There is abundant evidence that an estuary, as a unique ecosystem, is dependent upon the turbidity of the water it contains and that an estuarine ecosystem will be influenced by any change that occurs in the sedimentary regime.

Columbia River

Information on sediment transport and deposition in the Columbia River estuary has been obtained by measuring and sampling the flow, surveying the bed with acoustic techniques, and determining radionuclide levels. Flow measurements and water-sediment samples show that temporal and spatial discharges are large and are affected significantly by a turbidity maximum that develops and migrates longitudinally in the estuary. Side-scan sonar records obtained during a period of relatively low river flow indicate predominantly landward transport of sediment along the bottom in deep channels upstream from the mouth to about mile 14 and predominantly seaward transport on shallow slopes marginal to the channels downstream to about mile 5. A mass-balance equation that considers the amount of ^{65}Zn in the estuary bed and the net inflow of ^{65}Zn to the estuary suggests that approximately 30 percent of the silt and clay that enters the estuary from the river is retained there. Because of the complex character of estuary flows, detailed information on transport can be obtained only by making observations throughout tidal cycles and over extended periods of time (Hubbell and others, 1971).

Whatcom Waterway

Sediments from Whatcom Waterway, Bellingham were studied in response to a proposed dredging and disposal program. Laboratory study indicated that two types of sediment were involved. Sediment from the inner harbor consisted primarily of putrefying pulp fibers that exerted a significant oxygen demand, created substantial turbidity, and were toxic to juvenile sockeye salmon because of their hydrogen sulfide content. Various methods of widespread dispersal to dilute the sediment appeared impractical, and it was concluded that land disposal of inner harbor sediment would be necessary to protect fish stocks. Sediment from the outer harbor was a natural silt, not containing hydrogen sulfide, but exerted an oxygen demand and created a highly turbid mixture that settled very slowly. Because dumping of this sediment at the proposed site could also prove harmful to fisheries, hydraulic dredging and local disposal adjacent to the outer harbor was recommended (Servizi, 1971).

A remarkably small percentage, no more than 5 percent, of the particulate suspended matter in the Zaire River does in fact reach the open ocean. This material reaches the ocean as particulates that are less than 18m. Of this total population of particles suspended material having a particulate diameter of 9.5-18 m drop out onto the continental slope and the nearby ocean floor; to a smaller degree particulates having a size range of 5-9.5 m also fall out on the continental slope and on the nearby ocean floor. On the other hand, particles that are less than 5m in diameter by and large are held in suspension.

The bulk of the suspended particulate matter in the Zaire is apparently eliminated from the surface water by the time salinities are twenty parts per thousand or less. The particulates of larger diameter settle out simply. The fine particulates are removed primarily by a process of coagulation, scavenging, or both.

In the ocean proper the suspended particulate matter is made up almost exclusively of particles that are smaller than 10m. Somewhere around 50 percent of the total suspended particulate matter which is carried down to the sea by the river is settled out in the estuary "in the head of canyon; some settles out in the mangrove swamps on each side of the main channel of the river." Oceanward there is a relative increase of fine material and organic matter in the surface water. Water off the coast at the surface usually has a larger amount of suspended matter and of organic matter than does water taken from one meter below the surface.

Zaire Estuary

An informative study on the suspended matter in the Zaire (Congo) River has been published by Eisma et al. (1973). Their investigation shows that the Zaire River carries a comparatively low load of suspended particulate matter. The authors attribute this condition to "the predominance of chemical erosion and low gradients in the river basin." The suspended particulate matter is made up approximately as follows:

> Koalinite 26%
> Quartz .22%
> Organic Matter 32%
> Ironhydroxydes 10%

There are small amounts of other clay-type minerals and feldspar. Organic matter is carried by the river down to the sea grossly as floating vegetation and fibers from various plants.

The density range of the suspended particulate matter is apparently somewhere between 1 and 2.66.

The major elements in the suspended particulate matter in the surface waters of the Zaire River and Estuary have been made by Sholkovitz and others (1978). The suspended material seems to be enriched in aluminum, iron and phosphorus. It is depleted in silicon, calcium, potassium and magnesium, "relative to recent pelagic sediments and to suspended matter carried by the Amazon River." These modifications are characteristic of the soils found in regions where intense chemical weathering occurs. In western equatorial Africa such chemical weathering is common.

The authors found out that: "suspended samples from waters near the bottom of the estuary have compositions that indicate that this material has undergone much less chemical weathering. These particles were of greater size and most likely contained a larger proportion of quartz, micas, feldspars and oxides of iron and titanium than those of the surface waters" (Sholkovitz et al., 1978).

The trace element composition of suspended particulate matter in the Zaire Estuary is not particularly unusual. There seems to be a close agreement between the concentrations found in the Zaire River and those found in other rivers on the planet.

"Within the mixing zone, lanthanides Th, Sc, Hf and Rb remain nearly constant, whereas Cr, Co, Ni, Cu, Zn and Pb concentrations show much more scattering and are either constant or increasing" (Martin and others, 1978).

The elemental content of the suspended particulate matter in the Zaire River and in the mixing zone was ascertained by neutron activation analysis. The results are in Table 6-II (Martin, Thomas and Van Grieken, 1978).

Table 6-II.

ELEMENTAL CONTENT OF ZAIRE RIVER SUSPENDED PARTICULATES. VALUES PARTS PER MILLION DRY WEIGHT. NEUTRON ACTIVATION ANALYSIS.

Element	Zaire River	Zaire Mixing Zone
Ag	35.	7.4
As	3.8	-
Au	0.05	0.13
Br	6.4	37.
Ce	74.	76.
Co	30.	17.
Cr	211.	158.
Cs	5.3	3.7
Eu	1.7	1.9
Hf	5.4	4.3
La	53.	40.
Lu	0.36	0.31
Na	2100.	20000.
Sb	0.5	0.6
Sc	18.	16.
Sm	8.	7.
Ta	0.9	0.9
Tb	0.6	0.8
Th	16.	16.
Yb	3.	3.
Zn	300.	280.

Zaire Estuary, Iron

The true content of dissolved iron in river and estuarine water is much lower than has been reported in earlier literature (Figuères and others, 1978). If Zaire water samples were passed through filters of smaller porosity the particulate nature of much previously reported dissolved iron was illustrated. As water moving downstream in the river is mixed with seawater somewhere between 50 to 80 percent of the supposedly dissolved iron leaves solution by way of coagulation and precipitation on grains of silica. It is evident that "the use of filters with different pore size showed that the rate of removal [of iron] is directly related to pore size." (Figuères and others, 1978).

The concentration of iron in solution in the waters of the Zaire Estuary varies inversely with the pore size of the filters through which the water sample has been passed before analysis (Figuères and others, 1978). The concentration of the iron in the filtered water varies inversely with salinity.

For example if the pore size is 1.2m diameter, the variation of iron with salinity may be expressed:

$$\log Fe = 2.58 - 0.06 S$$

where Fe is given in microgram per liter and S, salinity, in parts per thousand. The curve is obviously not linear; $r = -0.92$, $r^2 = 0.85$.

Similarly, if the pore size is 0.05m diameter the relationship is as follows:

$$\log Fe = 2.11 - 0.04 S$$

where Fe is given in gm/l, S in 0/00, r is -0.98 and r^2 is 0.95.

Uranium

In the Zaire River there is a low concentration of uranium-238 in solution. According to Martin and others (1978) the dissolved uranium-238 "behaves as a conservative tracer during estuarine mixing." The amount of uranium dissolved in Zaire River water is much like the concentration found in the waters of the Amazon River. Martin and others (1978) maintain that "within the scatter of the ^{234}U - ^{238}U data there is most likely a preferential leaching of ^{234}U from suspended sediments during estuarine mixing due to the increasing concentration of complexing ions." This leaching process may very well deliver an additional amount of ^{234}U to the ocean.

The Zaire River estuary and the plume in the mixing zone have been studied in an orderly and thorough fashion by Dutch oceanographers under the sponsorship of the committee on oceanic research of the Netherlands Academy of Sciences. Financial support was insured by the Netherlands Ministry of Education and Science. The Zairean government cooperated and gave the scientific group permission to carry out detailed research in the territorial waters of Zaire. The results of this orderly, important and well-executed research endeavor have been published in the *Netherlands Journal of Sea Research*, Vol. 12(3/4), 1978. In essence this is a symposium volume with the title "Geobiochemical Investigations in the Zaire River, Estuary and Plume." The material published in 1978 covers pages 253 through 420 of the *Netherlands Journal of Sea Research*. The particular issue deserves careful study because of the excellence of the planning, the execution, and the analysis. It is also worth comment that the research program began in November, 1976, and the results were published in the December issue of the Journal, 1978.

Lagoons

Lagoons that are associated with various kinds of tropical and subtropical islands have water that is much different in composition from the adjacent open sea. Moreover, there is considerable difference with respect to the composition of lagoon water from one lagoon to another (Castanares and Phleger, 1969).

Lagoons with the estuarine type of water circulation have the maximal load of suspended particulate matter at the region which is about ten parts per mille salinity. In lagoons that show no difference in salinity from one part to another there is a gradual decrease in the concentration of total suspended particulate matter as one moves seaward.

As a generalization the percent of living organic matter increases as one moves from estuarine to neutral through anti-estuarine and finally into open sea regions. The total organic matter in suspension is virtually the same in the estuarine, the neutral, and the anti-estuarine regions; there is a dramatic decrease in total organic matter in the open sea.

Studies have shown that humic and fulvic acids have growth-promoting effects on unicellular organisms. Suspended matter in an estuary generally tends to increase significantly as the tidal velocity increases; the suspended particulate matter tends to fall off in concentration as tidal velocity decreases.

Amazon Estuary

Sholkovitz and Price (1980) made a detailed and important study of the chemical composition of suspended particulate matter in the Amazon Estuary. They come to the general following conclusion: "This study has demonstrated that the chemical composition of river-introduced suspended matter can be significantly altered by biological activity within estuarine waters as can be the geochemical cycle of inorganic elements."

Rhine River

Duinker and Nolting (1978) have shown that in the Rhine Estuary several trace metals are taken out of solution during the process of estuarine mixing; these metals are adsorbed by particulate matter. Some of this metal is deposited in the bottom sediments of the estuary. Such deposition results in "higher metal concentrations than can be accounted for by mixing of fluviatile and marine derived particles." The Rhine Estuary is seen to function as a sink for both dissolved and particulate trace metals. Only a small fraction of the total amount of several trace metals carried by the river actually reaches the ocean. These observations are in accord with those of other investigators who have shown the scrubbing out action that particulate matter in estuaries has for trace elements.

Fate of Metals in Estuaries

Turekian (1977) has published a valuable and succinct discussion of the role of particulate material in cleaning up the aqueous system of metals found in estuaries. The use of Pb_t provided a method for tracing pathways of metals through the system of a river-estuary-open ocean. The atmosphere and the process of weathering introduces into soils a variety of trace metals. These metals are carried by water from the land into flowing streams where they are concentrated on suspended particulate material. Many of these metallic elements are removed from the aqueous portion of the flowing stream in the form of particulate material at the boundary of the stream and the ocean into which it flows. A few elements are released. The cycle of trace metals in an estuary may be called "self-contained." Turekian (1977) maintains that "release and deposition occur virtually entirely within the system." It is further maintained that the major accumulation of trace metals in the oceans is carried out through suspended particulate matter. The overall end result is that oceans are remarkably free of metals. "The great particle conspiracy is active from land to sea to dominate the behavior of the dissolved species."

Chapter 7

GEOGRAPHIC STUDIES

General

VARIABILITY in the distribution of suspended particulate matter from place to place in the ocean has been illustrated by Gardner (1977). The variation of suspended particulates with depth at one oceanic station was:

$$y = 346 - 0.28x \quad r = 0.74$$

where y is the concentration of suspended particulates in micrograms per liter and x is the depth in thousands of meters.

At another station not far removed the variation was:

$$y = 15.2 + 0.02x \quad r = 0.75$$

Baffin Island

Gilbert (1978) described the pattern of suspended sediment in Pangnirtung Fiord, Baffin Island. The fiord has a low sill and a tidal range of close to 7 meters. Bottom water is replaced almost continuously. The concentration of suspended sediment in mg/1 in surface water of the fiord varies as follows:

Kilometers from head of fiord	Concentration
40	2
24	2
12	4
10	4*
6	10
4	18
2	22

*One measurement gave 16mg/1 on 5 August 1977.

Circulation is rapid in the fiord. The variation of suspended sediment concentration with depth is complex.

Greece

What turbidity changes are observed in areas of generally polluted water? Ignatiad and Becacosk (1970) observed such changes over a twelve-month period off Pireus Point, Greece. Water transparency decreased because of anthropogenic suspended particulate matter. Secchi disk values over the first eight months of the observation period clustered around 2 meters; during the latter part of the year of observation values of 2.5 to 3 meters were recorded. The average vertical extinction coefficient K was 0.5 in the generally polluted area; 0.3 in presumably clean areas in the open sea. Two interesting points were made by the authors: "... the fouling organisms can resist the toxicity of the oil polluted water" and "The existence of a great amount of suspended mud also did not seem to affect the filtering mechanism of the filter-feeding organisms."

Blue, green, and red light respectively were differentially absorbed in the polluted area:

Month	Approximate % of surface light at 3.5m		
	Blue	Green	Red
October	22	15	10
November	26	32	15
February	1	1	1
May	6	6	3

Beaufort Sea

Reimnitz and Barnes (1973) made a study of the suspended sediments of the Beaufort Sea. They used Secchi disk readings to obtain the light attenuation coefficient; particle weights were obtained from filtration. They found a remarkable difference over a two-year period. The distribution of turbid water was uniform, in August, along the north Alaskan coast. Discharge from the rivers was low at that time. It seemed to the authors appropriate to conclude that turbidity is related to factors other than river effluent. In their view the action of waves is more important than has been previously considered.

Bay of Bengal

The distribution of particulate organic matter in the Bay of Bengal was studied by Rao and Rao (1975). Monthly variations of particulate organic carbon, nitrogen, and phosphorus respectively showed similar trends. The values follow in g at/l:

Element	March-April	October-November
C	14-15	8-13
N	1-1.3	1.2-1.9
P	0.2	0.1

The March-April period is the time of peak growth of phytoplankton. In July and August along the Waltair Coast of the Bay there is heavy rainfall; the concentrations of C_{org} and N_{org} increase.

The overall concentration of organic matter in particulate form is comparatively low in the Bay.

Mid Atlantic Bight

Studies in 1969 of the Middle Atlantic Bight (coast between Cape Cod and Cape Hatteras) indicate that suspended matter collected there was mostly organic. Surface waters contained 80 percent combustible matter; near bottom waters, 40 percent. Total suspended solid concentration decreased between the inner shelf and the shelf break by an order of magnitude in near surface and near bottom water. Ash fraction of suspended matter decreased over the same distance by one order of magnitude in the near bottom water and two orders of magnitude in near surface water.

"Recently contributed river sediment is not a significant constituent of the suspended matter in the waters of the shelf, particularly the outer shelf. Most of the inorganic material in suspension represents resuspended bottom sediments (at least some of which are relict) whose suspended concentrations are increased notably by storms" (Meade and others, 1975).*

The Bering Sea

Regional variations in particulate matter suspended in the ocean have been addressed by Japanese investigators in cruises in the Bering Sea (Nakajima, 1969). During a cruise of the Oshoro-Maru a variety of measurements were made of seawater samples. These included temperature, salinity, oxygen, nitrate, total dried weight of seston, particulate organic carbon and nitrogen, chlorophyll *a* and pheopigments. In general there were high concentrations of particulate matter in water near the surface of the sea. With increasing depths there was a rapid decrease in concentration of particulate matter down to around 50 meters. The biomass of phytoplankton as estimated from chlorophyll *a* increased "more or less rapidly" with increasing depth. "This fact suggests the possibility that an intensive production of particulate matter occurs at the sea surface through although the mechanism of surface production is not clear" (Nakajima, 1969).

There was demonstrated an inverse relation between chlorophyll *a* and the suspended particulate matter in the surface waters at every station sampled in the Bering Sea. A clearly defined minimum was documented just below the thermocline. The precise mechanism for the formation of this minimum layer is not clear.

Three characteristic maxima in deep water layers below the thermocline were identified. "These were considered to have resulted from 'renewal' of particulate matter." Horizontal advection is presented by Nakajima (1969) as the most probable explanation.

It is finally concluded that: "The results obtained in this cruise support the concept that the vertical zonation of particulate matter is well associated with the stratification of water mass..."

Nakajima and Nishizawa (1972) have shown that the amount of particulate organic carbon in the surface waters of the Bering

---

*Meade (1972) has pointed out that on the Atlantic Continental Shelf of the United States river sediments are a less important source of suspended material than is biogenic detritus coming from organisms of the shelf and material that is resuspended from the shelf bottom. Suspended material accumulates chiefly in the large estuaries and marshes of the coast.

Sea decreases with depth quickly in depth intervals that are less than 100 meters. The trend of this rapid decrease in concentration is almost precisely exponential. One-third of this decreased concentration may result from the use of the particulate organic carbon by ultra-plankton and bacteria. "The average sinking velocity of these particles was estimated to be close to 1m/day" (Nakajima and Nishizawa, 1972).

Nishizawa and Nakajima (1971) have reported on the amount of particulate organic material in the surface layers of the waters of the Bering Sea and compared these findings with similar concentrations in the subtropical and tropical areas of the north Pacific Ocean along 142° east meridian line. At or close to the surface of the ocean a maximum concentration of particulate organic material amounting to 50-200 micrograms of carbon per liter was recorded. The concentration showed a rapid drop with increasing depth in the stratum from 50-100 meters. There was a tendency to level out at a minimum concentration amounting to less than half of the value at the surface. "This pattern was in sharp contrast to that of chlorophyll, which almost always was characterized by a more or less distinct subsurface maximum occurring near the base of the euphotic layer." It was estimated that the rate of particulate formation at the surface of the ocean is about 260 milligrams of carbon per square meter per day.

Pacific Ocean

In the Pacific Ocean, Soviet oceanographers have observed variations in water transparency near the bottom of the sea (Kozlyaninov, 1974). In the Isu-Bonin Trench, as an example, the extinction coefficient near the bottom (0.75-0.80 m) was virtually ten times that measured at the surface. The explanation of these observations lies in "the possibility of mudslides on its slopes, i.e. the Isu-Bonin Trench, with the resulting sharp increase in the number of particles suspended in the water."

North Pacific Ocean

Nakajima (1973) has reported the results of an extensive study of suspended particulate matter in ocean water sampled down to 4,000 meters in the north Pacific Ocean.

Sixty of the collecting stations were located at a line 142° E. The stations were north of the equator up to 30° N. Additional stations in the subtropical ocean were added: 7 on a line 125° E; 2 on a line 132° E; 3 on a line 125° E (east China Sea). Additional stations in the western north Pacific Ocean and adjacent seas were surveyed.

Dried weight of particulate matter exhibited a regular picture of vertical distribution throughout the area studied.

The average amount of particulate matter in the upper 0-50 meter layer "was linearly correlated with the average concentrations in deeper layers of the same area." Nakajima (1973) contends that there is a pronounced "regionality" with respect to the amount of particulate matter in the entire column of water under study. Furthermore "the regional variation is ultimately controlled by the regional variation in primary production in the surface photic zone."

An idea of the values obtained and the relationships of total seston dried weight, percent carbon fraction, and percent nitrogen fraction are explained as follows. In the area from 19°N to 28°N along line 142°E the mean seston dried weight in milligrams per liter in the 0-50 meter layer was 0.24; in the 2,000-2,500 meter layer, 0.16. The relationship of seston concentration to depth may be expressed as follows:

$$y = 0.23 - .00011x$$

where y is the calculated total seston dry weight in milligram per liter and x is the depth layer in meters.

The relationship of percent carbon to total seston dry weight can be roughly expressed by the equation:

$$y = 26.5 - 64x$$

where y is carbon in percent and x is seston dried weight in milligrams per liter.

In the Kuroshio area the carbon fraction of the seston varies with depth as follows:

$$y = 17.7 - 0.01x$$

where y is the calculated percent carbon in the seston and x is the depth layer in meters.

Average carbon fractions as calculated from the data collected were never greater than 25 percent.

According to Nakajima (1973) the inverse relationship between seston dried weight and carbon content "means that the organic fraction is minor compared to the inorganic fraction. Average carbon fractions calculated for each depth range ... never exceeded 25 percent."

Table 7-I summarizes some of the analytical results reported by Nakajima (1973).

Table 7-I.

AVERAGE CHEMICAL COMPOSITION OF ZAIRE, AMAZON, AND RIO NEGRO RIVER SUSPENDED MATTER. WEIGHT PERCENTAGE. THE AVERAGE FOR THE 3 RIVERS IS COMPARED TO THAT FOR CENTRAL ATLANTIC PARTICULATES AND TO THE AVERAGE COMPOSITION OF MARINE SEDIMENTS.

Chemical	$\bar{X}$	S	S^2	Central Atlantic	Marine Sediments
SiO_2	51.7	2.8	7.8	66	60
TiO_2	0.8	0.033	0.001	1	1
Al_2O_3	29.6	3.1	9.6	18	19
Fe_2O_3	13.0	1.2	1.3	8	9
MnO	0.13	0.03	—	0.2	0.5
CaO	0.81	0.015	—	0.7	1
MgO	1.2	0.26	.07	2.7	4
Na_2O	—	—	—	1	1.7
K_2O	1.4	0.74	0.55	2.2	3.6
P_2O_5	0.4	0.13	0.02	0.1	0.2

Table 7-I, p. 409, data reworked (Nakajima, 1973).

Sugihara and Tsuda (1979a & b) have studied the volume-scattering function and size distribution of suspended particles in the North Pacific Ocean. Scattering coefficient was estimated from the observed volume scattering function; cross-section concentration of particles > 2.4m diameter was measured. A linear relationship was ascertained between these two functions; this relationship held for northern and southern regions of the central North Pacific. The constant of proportionality between northern and southern regions was greatly different. Indeed "the constant in the southern region was too large compared with the values obtained by the Mie theory." This anomaly was traced to the neglecting of particles > 2.4m for technical reasons. The refractive index of particles was obtained from the correlation between the scattering coefficient and the volume scattering functions. In the North Pacific the index was found to be 1.03-1.05. In Tateyamo Harbor (very turbid water) the value was the same. For river water entering the Harbor the values were 1.1-1.2.

A wedge-type interference filter and grating was used to measure spectral energy distribution of underwater light off the coast of Japan (near Muroran and Hakodate respectively) and in north Pacific waters at 45°00'N by 164°58'W (Sugihara and Inove, 1976). Color change with depth was ascertained for each area. The coastal waters around Holskaido Island had the optical characteristics of very clear ocean water. Water from the station in the North Pacific Ocean was turbid.

Sugihara (1977) published an interesting mathematical treatment of the underwater light field. He calculated the radiance distribution brought about by scattered light, using scattering up to the third order. He found by his method that "multiply scattered light plays a more important role as the depth increases." Computed and observed radiance values agreed rather well in the vertical distribution but not in the angular. The discrepancy could result from a deficient attenuation coefficient and volume scattering function used in the calculations. A table of estimated optical properties of seawater at various depths is included for depths from 25 to 300 meters.

Sugihara and Tsuda (1979a, b) used a Coulter Counter to ascertain the size distribution of suspended particles in the central North Pacific Ocean from about 45°-22°N by about 160° to 165°W. Particulate concentration was high in the north portion of the study region; it decreased rapidly as one moved southward. Cumulative distribution of the particles could be described by a hyperbole. In the north particles of less than 4.8m in diameter were distributed in a fashion that fit both a hyperbolic or an exponential form. In the south these particles were distributed hyperbolically. "Thus, the difference of the shape in size distribution rather clearly existed between the northern and southern regions."

Particles with diameters ranging from 3.8m to 9.6m made up a larger fraction of volume in the north than in the south. The relative volume of particles of different size at about 32°N about 4 percent for those from 2.5m to 9m. Above that diameter there was a rapid rise to a peak of 30 percent at a 15m diameter followed by a fall to about 11 percent for particles of 22m diameter.

The Sargasso Sea

In the Sargasso Sea the fall-out of suspended particulate matter is said to be comparatively slight. Teal and Teal (1975) estimate the fall-out to result in the accumulation of one millimeter of sediment on the Sea floor every 1,000 years. This situation is of course unique to the Sargasso Sea. The value of one millimeter of sediment deposited on the Sea floor every thousand years must not be considered as representative of the ocean floor in general or indeed of the ocean floor of any extensive part of the world ocean.

The water of the Sargasso Sea may be termed very clear ocean water; its suspended particulate load is characteristically 0.05mg/l or less (GESAMP, 1975).

Honjo (1980) has published a detailed study of the flux of particulates in the central Sargasso Sea, the tropical Atlantic, and the central North Atlantic. This paper includes important experimental and observational information.

African Coast

The surface layer in a northwestern African coastal upwelling holds 0.2-0.4 mg particulate/liter.

These values are of interest in comparison to estimated values to be found in the top 100 meter layer of surface water: immediately upon disturbance of the bottom the suspended matter will rise to a maximum of 2.5mg/l; after 10 hours various diluting factors will bring the maximal concentration of suspended matter to about 0.05-0.01 mg/l as a result of deep sea mining (Kullenberg, 1975). Judicious use of the natural diluting factor could keep surface turbidity at a very low level during mining and similar activities.

Africa

Emery et al. (1974) made a study of suspended particulate matter from 1534 stations along the west coast of Africa for a period of two years. The most concentrated total organic and inorganic suspended particulate matter was found above the continental shelf north of latitude 7°N. A high concentration of suspended organic particulate matter seems to be associated with coastal water upwelling. High concentrations of inorganic suspended particulate matter along the coasts originate from major river inflow and from heavy dust falls.

If values for total suspended matter in milligrams per liter are plotted against the measured Forel color, percent yellow, there is considerable scatter but a curvilinear correlation obtains (log X log plot):

Average Suspended Matter, Milligrams Per Liter	Forel Color, Percent Yellow
0.1	2
0.15	4
0.2	10
0.3	20
0.7	30
1.0	60

The general values for suspended solids arranged in 1972 from 0.08-1.0 milligram per liter. In the coastal zone between Guinea and Lizman the average value was 0.5mg per l. The overall general average was 0.15mg per l.

Offshore Transport of Sediment

"Turbidity currents and/or sliding can be very effective means of transporting material offshore in areas of some slope, say 2-3° or more. The possibility of triggering sliding motion or even small turbidity currents artificially should be borne in mind. This is perhaps in particular true for activities along the shelf break or slope, i.e. in connection with offshore mining" (Kullenberg, 1975).

Neumann and Pierson (1966) have examined the turbidity current phenomenon and conclude that the majority of marine geologists see turbidity currents as potent mechanisms for eroding submarine canyons and flattening out abyssal plains.

The role of turbidity currents and their geological effectiveness are still under active study and debate. It is appropriate, however, to suggest that, whenever large-scale sediment movements in the ocean are under consideration, turbidity currents must be given adequate attention and close study.

"Offshore transport in the breaker zone occurs as both suspended and bed load. Outside this zone, in the offshore zone, the sediment transport is mainly bed load, the energy dissipation being caused mainly by bottom friction" (Kullenberg, 1975).

According to Hails (1974) information about transport of suspended sediment by wave action is, by and large, limited to the top few meters of water.

Influx of Oceanic Suspended Material

How much suspended material is introduced into the ocean each year? Precise, quantitative information is lacking; reasonable estimates have been made (Waldichuk, 1977).

Materials can get into the ocean from land via river discharge, by transport through the atmosphere, and then washing down to the ocean by rain, or by direct injection into the ocean by manmade outfalls. About 180×10^{14} grams per year of suspended solids enter the oceans by way of rivers. The effect is ordinarily along the coastal margins. "Oceanic dispersion processes are just not rapid enough to distribute river-borne materials globally within the time frame we are accustomed to in the atmosphere (Waldichuk, 1977).

Atmospheric transport of materials from land to ocean is becoming ever more obvious as an important and relatively rapid route.

It is important to recognize that the coastal waters of the world make up about 10 percent of the surface area of the world ocean, "but 99 percent of the world fish catch originates from these coastal waters and from relatively small (0.1%) upwelling areas. Most of the open ocean, lacking in nutrients to sustain life, is a biological desert."

The Mediterranean Sea is located in such a manner as to be surrounded by a large human population. Ninety-nine percent of the border of the Mediterranean Sea is made up of closed coastline. It is estimated that around 100 million people live in the coastal areas surrounding the Mediterranean Sea. By the year 2000 it is estimated that this population will have doubled.

Man-made activities of various sorts bring into the Mediterranean a significant load of suspended material, both organic and inorganic.

Japan — Bristol Bay

Japanese investigators have shown great interest in the optical properties of sea water. Sugihara and Kawana (1975) measured the intensity of polarized light components in turbid sea water collected from Bristol Bay. Horizontally and vertically polarized components showed a small difference for a measured volume scattering function if the scattering angle were small. As the scattering angle increased there was a rapid decrease of both components; the horizontal component decreased faster. Degree of polarization roughly followed a symmetrical curve if scattering angle was on the x-axis. The polarization was greatest at 90° angle and sloped downward on each side of 90°; but, "the value at 135° is a little higher than that at 45°." Several assumptions were made about the suspended particles: refractive index = 1.20; are in the form of nonabsorbing spheres; a certain size distribution. Using the Mie theory, the degree of polarization and the volume scattering function were calculated. Poor agreement between theory and observation obtained. The authors maintain: "Judging from the computed results of the degree of polarization as a function of size in case of a single sphere, this discrepancy is considered to be due to the effort of particles larger than 2, under the assumption that the refractive index of particles is equal to 1.20."

Mediterranean Pollution

In the following comments the suspended matter will be given in terms of total suspended solids times 10^6. Domestic pollution that originates from the coastal zone of the Mediterranean brings in 0.6 tons per year. From industrial sources, the amount of total suspended solids is 2.8 X 10⁶. Agriculture is a large contributor which amounts to 50 tons per year. It is obvious then that the total pollution loads originating in the coastal zone insofar as suspended matter of concerned equals 53 X 10 tons per year. Rivers carry a natural, or so-called normal, load of suspended solids into the Mediterranean. The pollution load of suspended matter cánnot be estimated simply because there are insufficient data on which to base an estimate. The background load is about 300 X 10 tons per year carried into the Mediterranean by rivers. The total load of suspended matter carried into the Mediterranean by all sources that have been measured equals about 350 X 10⁶ tons. The range is 100-600 X 10⁶ tons per year (Juda, 1979).

The Mediterranean Sea illustrates what seems to be happening inevitably in other bodies of water surrounded by ever-increasing populations of mankind.

Canada

Bartlett (1977) has studied the deposition of suspended particulate matter in the coastal zone of Prince Edward Island, Canada (1977).

Measurements indicate that suspended particulate matter at the headwaters or the upper estuary regions amounts to 180 milligrams per liter. In the lower estuary or lagoonal regime the amount of suspended particulate matter is about three milligrams per liter. In general, the values increase in March and April; this is the time of spring runoff. The general increase of suspended particulate matter at the air-sea surface is about four mg per l; at the sediment water interface it is about 80 mg per l.

"The resultant turbidity plays an important role in limiting photosynthesis in the water mass and in the concentration of oxygen, carbon dioxide, and specific heavy metals and nutrients in the bottom sediments" (Bartlett, 1977).

For Saguenay Fiord see Chanut and Poulet (1979).

Panama

Plank et al., (1973) made a study of suspended matter in the Panama Basin. They used light scattering methods and Coulter counter measurements on samples of seawater taken at fifty different collecting stations. The data they collected indicate that there are three possible sources of suspended particulate matter in the Basin. First of all, there are the surface waters throughout the Basin; secondly, erosion and run-off from the continents; and finally, erosion of the bottom at two collecting sites on the Carnegie Ridge. The distribution of the suspended particulate matter was generally in accord with the usual pattern of abyssal circulation.

Siberia

The USS Burton Island during a cruise in the east Siberian Sea collected turbidimetric measurements along with other types of oceanographic information. The actual measurement of turbidity in water samples taken during this trip was made using Hellige turbidimeter. The primary purpose of these studies was to obtain data on the time and spatial stability of particulate material in the sea at various locations (Abbott and Hiller, 1966). There are other kinds of data that bear out evidence that turbidity layers are also sound-scattering layers and may play a basic role in acoustic volume reverberations.

The instrument used for these measurements was the so-called low-range Hellige turbidimeter; its reference range was 0-150 ppm SO_2. The instrument makes use of a principle based on the Tyndall effect in which light passing upward through a sample of water is matched against a light from the side scattered upward by the suspended particles in the water. When the variable center lighting matches the outer portion of the sample, the dial number on the side of the instrument is read directly. This number, which has no dimension, is then used to arrive at the turbidity of the sample by referring to a prepared calibration chart for that specific instrument and equipment. The average turbidity figure obtained during this cruise was about five parts per million. This value seems rather high if it is compared to turbidities of about 0.5 parts per million or less in portions of the Atlantic Ocean that have been tested. Visual inspection of the ice and the water gave the impression that both were rather dirty. If a significant amount of this particulate material were discharged into the water it might explain the high values reported. The authors also suggest that "Another possible cause may have been the relatively shallow depth of the water in which the stations were taken, the average depth of all stations reported in this area being about 30 meters." At several of the sampling stations repeated measurements were made over a period of time of about 48 hours. At these stations there were some rather interesting variations in turbidity from time to time. A number of explanations are offered such as turbidity currents, tidal currents, wind, and the like.

Australia

The spectral distribution of radiation in southeastern Australian waters has been described (Kirk, 1979).

Spanish Coast

Fraga (1979) made observations on the relationship of Secchi disk visibility to concentrations of phytoplankton and of suspended clay in the water. The study was part of an overall investigation to ascertain the effect of water turbidity on primary production in the Vigo Estuary in Spain.

The equation relating these values, using the least squares method, is:

$$1/S = 0.067 + 0.0116\, \text{Chl}\, a + 0.128\, \text{PMM}$$

where PMM is particulate mineral matter in mg/l; S is Secchi disk visibility in meters; Chl a in gm/l.

Other values of interest were reported:

Total attenuation coefficient, 1.7/S

Mean absorbance of clear water, 0.11/m

Attenuation coefficient due to phytoplankton, 0.028/m

Attenuation coefficient due to clay, 0.21/m

Photosynthesis in a column of sea water is said to be an inverse function of the water attenuation coefficient (Fraga. 1979). Attenuation coefficient equals 1/S, with S the Secchi disk reading in meters. Suspended clay has been shown to depress productivity by a factor of 26 percent.

Fraga and Mourino (1978) have published detailed hydrographic data for "ria de Vigo" for the years 1975 through 1977. Turbidity data are included. The trend line for Secchi disk readings over the 1975-1976 12-month period of observations shows a gradual increase in values from May 1975 until about mid-January 1976 as follows:

$$y = 5.98 + 0.37x$$

where y is the Secchi disk reading in meters and x is the month. (May 1975 is month zero; June 1975 is month 1; May 1976 is month 12.)

About mid-January there is a sharp downward break in the trend. From that time until May 1976 (end of observations) the trend of Secchi disk readings is downward:

$$y = 19.54 - 1.2x$$

where the symbols are the same as in the previous equation.

During the period October 1976 through December 1977 there were irregular variations of Secchi disk values: highs were found in November '76, April, September and December '77. Extreme lows (turbid water) were recorded in early December 1976, late July 1977 and mid-December 1977. The trend line over this period was virtually horizontal.

North Sea

Eisma and Gieskes (1977) made a study of the particle size distribution of nonliving suspended matter in the southern North Sea. Essentially they found that "nonliving particles in nearshore (coastal) waters were found to be on the average somewhat finer and better sorted than in the offshore (channel) water" (see also Eisma and Kalf, 1979).

Sognefjord

Eisma and Kalf (1978) have made a study of the suspended particulate matter found in the Sognefjord between Norway and Shetland.

Orinoco Delta

Eisma et al. (1978) made a study of turbidity and bottom deposits around the Orinoco Delta. The results for turbidity measurements resembled reasonably well what has been described in numerous temperate zone estuaries.

Oregon Coast

McCave made a study of the suspended particulate matter over the central Oregon continental shelf during the month of May, 1974. The concentrations of organic and of inorganic fractions were recorded and the fluxes of the different components mapped out in reasonable detail. The author summarizes his findings in a diagram which is most useful in understanding the various interrelationships. The picture that is seen from this research is conventional for areas of this sort. One exception seems to be the low percentage of labile organic matter recorded just above the lower thermocline where the water entering this shelf system comes from the Columbia River and contains a rather high percentage of organic matter.

Sohm Abyss

Amos and Gerard (1979) have identified a highly turbid water at the bottom of the ocean along the margin of the Sohm Abyssal Plain. The concentration of suspended particulate matter in this bottom water is reported to be one or two orders of magnitude higher than what is usually found in the waters of the deep oceans, i.e. 1500 to 4000 micrograms per liter. The investigators suggest that a turbidity current is the causative agent that carries this water downslope.

Texas Coast

Shideler (1979) has discussed the regional surface turbidity on the south Texas continental shelf during an eighteen-month period from November 1975 through May 1977. According to Shideler, "The overall shelf turbidity patterns reflect the degree of lateral interchange between the gulfward movement of turbid inner-shelf waters and the shelfward incursion of clear open-ocean waters."

Barents Sea

Atlantic Ocean waters exhibit less scatterance than adjoining Barents Sea waters. Exceptions occur if there are influences from primary production or particles rising from the bottom. The scattering coefficient of Atlantic waters is less than 0.1/m. Fluorescence in Atlantic and Polar waters is weak. Taking the exciting light at 366nm fluorescence at 525nm relative to Raman scatterance is about 4×10^{-3}/nm.*

Coastal Japan

Taguchi and Nakajima (1971) made a study of suspended particulate matter in several inlets from the ocean along coastal Japan. They found that the particulate organic matter was especially concentrated at the surface skin of the water studied. According to their results the maximal concentration factor was 17.6.

North Sea and Contiguous Waters

The suspended matter in the North Sea and contiguous waters has been described in some detail by Eisma (1980). Suspended particulate matter gets into the North Sea from the North Atlantic Ocean from the English Channel, from the Baltic and from a variety of rivers. The processes of coastal erosion, erosion of the sea shore and primary production also add suspended particulate matter to the water of the North Sea. Some also falls out of the atmosphere. Most of the suspended particulate matter seems to be concentrated near the coasts of the southern North Sea.

The amount of suspended particulate matter which gets into the North Sea is estimated to be about 34×10^6 tons per year.

Gulf of California

Zeitzschel (1970) has described the distribution and composition of suspended particulate matter in the Gulf of California. A Coulter counter was used and was found to be extremely useful in ascertaining the size, distribution and volume of particulate suspended matter in the ocean.

---

*Scatterance is the ratio of the radiant flux, scattered from a beam of light, to the incident flux.

Taiwan

Chan and Chen (1977) have described the composition of suspended particulate matter in the waters off Taiwan. The major mineral components found in the suspended solid material were quartz, feldspar, calcite, chlorite, caolinite and illite. They interpret the meaning of these various components in terms of origin of the water masses concerned. They conclude that: "The distribution pattern and the properties of the particulate matter are in accordance with the turbidity plumes and the upwelling zone which have been observed in satellite photographs."

As part of an overall study of the water characteristics of the Taiwan shelf and slope, description of the turbidity of the waters was made (Boggs et al. 1979). The turbidity of surface water in that region varies significantly in the summer and winter periods; the factors that seem to influence turbidity are changes in the weather, the winds and the ocean currents.

Helgoland

An important observational study was made by Luening and Dring (1979) in which they measured underwater radiance at intervals of 20 minutes for a period of one year. Measurements were made at depths 2.5 and 3.5 meters respectively. Half of the year the water is very turbid; but in the other half of the year, April to September, water is significantly less turbid. Information of this sort is of great value to marine biologists studying various aspects of oceanic and coastal productivity.

FRESHWATER GEOGRAPHIC STUDIES

South Africa

It is reported that all major rivers in South Africa have been modified by impoundments. The general result is a change in recurrence and magnitude of floods as well as stabilization of flow (Davies, 1979). These changes have resulted in known substantial environmental disturbances remotely located from the origin, e.g. Ben-Tuvia, 1973; Liebman, 1935; Oren and Hornung, 1972; Davies, 1979.

Norway

Studies of turbid water effects on salmon Norwegian streams seem to "affect only salmon angling in the river below (due to reduced transparency)." Reproduction of the salmon was not affected by turbid waters (Lillehammer and Saltviet, 1979).

Chapter 8

WASTE DISPOSAL, FRESH WATERS, ABATEMENT, LITERATURE REVIEWS

Ocean Disposal

THE trend of ocean disposal of wastes that are potential sources of turbidity in the years 1973-1979 shows a roughly linear decrease with time:

$$y = 10.69 - 0.43X, r = 0.76,$$

where y is the million tons of total waste disposed in the ocean for a given year and X is the year in the period 1973-1979, e.g. 1973 is year 1, etc.

How much suspended material is introduced into the ocean each year? Precise, quantitative information is lacking; reasonable estimates have been made (Waldichuk, 1977). (Table 8-I).

Materials can get into the ocean from land via river discharge, by transport through the atmosphere and then washing down to the ocean by rain, or by direct injection into the ocean by man-made outfalls.

About 180×10^{14} grams per year of suspended solids enter the oceans by way of rivers. The effect is ordinarily along the coastal margins. "Oceanic dispersion processes are just not rapid enough to distribute river-borne materials globally within the time frame we are accustomed to in the atmosphere" (Waldichuk, 1977).

Table 8-I.
U.S. OCEAN DISPOSAL OF WASTES, 1973-79*
(tons)

Type	1973	1974	1975	1976	1977	1978	1979
Industrial wastes	5,050,800	4,579,700	3,441,900	2,733,500	1,843,800	2,548,173	2,577,000
Sewage sludge	4,808,900	5,010,000	5,039,600	5,270,900	5,134,000	5,535,000	5,932,000
Construction and demolition debris	973,700	770,400	395,900	314,600	379,000	241,000	107,000
Solid wastes	240	200	0	0	100	0	998
Explosives	0	0	0	0	0	0	0
Incinerated Wood	10,800	15,000	6,200	8,700	15,100	18,000	36,000
Chemicals	0	12,300	4,100	0	29,700	0	0
Total	10,844,440	10,388,400	8,887,700	8,327,700	7,401,700	8,342,173	8,652,998

*Includes all dumping except dredge spoil, which represents 80-90 percent of all materials dumped.
Source: U.S. Environmental Protection Agency, Ocean Dumping, 1979 Annual Report to Congress (Washington, D.C. : U.S. Government Printing Office, 1980).

Atmospheric transport of materials from land to ocean is becoming ever more obvious as an important and relatively rapid route.

It is important to recognize that the coastal waters of the world make up about 10 percent of the surface area of the world ocean. "But 99 percent of the world fish catch originates from these coastal waters and from relatively small (0.1 percent) upwelling areas. Most of the open ocean, lacking in nutrients to sustain life, is a biological desert" (Waldichuk, 1977).

The volume of dredged material taken from United States harbors and waterways and dumped into the ocean has varied remarkably since 1973 as the following table shows (Council on Environmental Quality, 1981):

Year	Million Cubic Yards of Dredged Material Dumped
1973	67
1974	99
1975	88
1976	66
1977	41
1978	53
1979	73

Obvious biological or ecological effects, if any, of this activity are not documented in the conventional scientific literature.

Control of Man-made Turbidity

Silt curtains have been evaluated as effective barriers to the spread of surface silt from marine dredging and construction activities (Anon., 1979). These curtains are floating obstacles or baffles anchored so as to form a U-shaped or completely circular form as viewed from above. The greatest amount of the material discharged from the underwater operation creates a fluid mud at the bottom; it moves outward radially from the center of the dredging or construction operation. When the mud arrives at the bottom of the curtain, the curtain must be moved; otherwise it will be drawn underwater. Suspended matter in the water moves under the curtain. This effect significantly reduces the turbidity, resulting from the operation, in the upper part of the water column. In the presence of currents flowing at a high rate, the curtain flares outward either causing fluid mud to become resuspended, the layer of turbid water to appear at the surface just beyond the curtain, or both. A current velocity of 1 knot is reported to be the "limiting condition." The ideal clearance between the lower border of the curtain and the sea bottom, measured at low tide, is about one-half meter. The exact form of the silt curtain to be used will vary with configuration of the site of operation, the specific application, and the hydrodynamic circumstances. The material of the curtain requires a tensile strength of 525N/m, fabric weight of 610-746 g/m^2, tear strength of 445-890 N, tensile strength after abrasion 350 N/m. A bouyancy ratio of > 5 can be provided by appropriate solid, closed-cell plastic foam flotation material. "For currents > 0.1kn aluminum extruded load-transfer connectors are recommended."

Esthetics of Turbidity

Turbidity, and optical characteristics of water containing solids in suspension, influences human visual perception. The clarity of natural water is infrequently noticed alone; rather it is part of the whole visual field or scene. Most people probably concur that a clear mountain stream in an alpine setting is enjoyable whereas a turbid stream in the same landscape is unpleasant. Nevertheless the grandeur of the Green River in Utah crashing over great rapids is greatly intensified when the river is carrying silt. As a rule, however, elevated turbidities are interpreted as not pleasant. For example a turbid artificial impoundment (a reservoir) deemed unpleasing in the esthetic sense may attract fewer anglers. Turbidity created by gravel dredging in a river may be viewed as esthetically unpleasing for several miles below the dredge site.

Sparse research on the esthetic consequences of turbidity in natural waters has been completed. Methods for quantitating esthetic factors are primitive at best. Published reports of esthetic evaluations that include turbidity as a factor in the rating of water quality contain essentially no data that derive from the opinion of users or the public (see Sorensen et al., 1977).

By and large, attempts to express esthetic values either quantitatively or semi-quantitatively have not been successful. Evaluations of "esthetic qualities" invariably end up with the evaluator imposing his completely subjective views of what is beautiful on others. The situation merely reinforces the wisdom of the ancient adage, "De gustibus non disputandum est."

Russian Reviews

Russian scientists are in the process of producing a number of monographs that cover various aspects of what can be included under the general title of Oceanology. In a recent one the hydrophysics of the ocean is discussed (Kamenkovich and Monin, 1978). Chapter six in that monograph is rather lengthy (more than 50 pages) and devoted to discussing the optical properties of the ocean. Most of the chapter encompasses specific optical characteristics of seawater and reports on light fields in the ocean. A rather brief section is devoted to optical methods in oceanology.

Fresh Water Quality Guides

Wilber (1969) discussed at length the problem of turbidity and suspended particulate matter primarily as related to fresh waters. He concluded that streams may be usefully classified into several categories with respect to the amount of their suspended solids:

Class and Description	Suspended Matter (ppm)
I Optimal	25 - 30
II Good	30 - 85
III Poor	85 - 400
IV Extremely Bad	400+

In general excessive amounts of suspended particulate matter harm fresh water aquatic life in several ways:
1. Direct action on fishes or other organisms swimming in the water: killing them; reducing the rate of growth, decreasing resistance to disease and similar actions.

2. Prevention of successful development of eggs and larvae.
3. Modification of natural movements and migrations of organisms.
4. Reduction in abundance of food available to aquatic organisms.

Nitrogen and phosphorus availability in the Great Lakes is in part associated with the transport of suspended particulates from tributary rivers into the Great Lakes (Armstrong and others, 1979). Certain trace metals seem to be similarly associated with suspended particulate matter.

Intensity of Turbidity Research

The amount of research being done during the past decade on turbidity in the ocean is best indicated by quick reference to the large number of studies that have been published on various aspects of turbidity and its effects on living organisms.

Concerns have been expressed with respect to ocean dumping in the Gulf of Mexico, and turbidity is indicated as one possible measure of effect (Atlas et al. 1980). Regional studies of copper, lead, and zinc in the Antarctic Ocean have been published (Harris and Fabris, 1979). The behavior of many trace metals in different estuaries has been evaluated. The behavior of manganese in the Rhine Estuary has been described as related to turbidity by Wollast et al. 1979. Analytical methods for measuring many elements in suspended particulate matter using neutron activation analysis have been described (Ellis et al. 1979).

Yangawa et al. (1978) have discussed turbidity as related to commercial fishing. The amino acids found in suspended particulate matter in the Pacific Ocean have been characterized (Siezen and Mague, 1978).

The suspended matter in the ocean off Brazil has been discussed (Barretto and Summerhayes, 1975; Milliman et al. 1975).

Different approaches to studies of turbidity in the ocean and various kinds of observations have been included in the overall collection of turbidity reports (Kuroki, 1975; Proni et al. 1975; Pustel'nikov, 1974; Emery et al. 1974).

A few studies have involved the relationship of oceanic turbidity to the distribution of marine fish (Miller, 1974).

Japanese and Russian oceanographers both have been concerned with particulate suspended matter in the Pacific Ocean (Nakajima, 1973; Artem'ev, 1973).

There has been a gratifying expansion of research on oceanic turbidity since 1964 when Wilber (1964) devoted only about 3 or 4 paragraphs to it in a summary of information on animals in the aquatic environment.

In 1971 it seemed adequate to summarize information on the general aspects of oceanic turbidity in nine pages (Wilber, 1971a). A discussion of turbidity effects on marine animals required another nine pages (Wilber, 1971b).

Dynamic Aspects of Turbidity

During the decade of 1970 to 1980 there has been an astonishing increase in the number of published papers dealing in whole or part with various aspects of turbidity. The active role of suspended particulate matter (organic and inorganic) in the dynamics of estuaries (for example the role of particulates in the removal of river borne metals from water that ends up in the open ocean) is coming into focus. Turbidity as a factor in characterizing the nature and extent of discrete water masses seems clear, especially from the results of satellite observations. The origins of turbid waters are being disclosed by careful and novel investigations. It is not rash to suggest that turbidity phenomena will now rate a full chapter in modern textbooks of oceanography, rather than a paragraph or two.

Fresh Water Data

Fresh water fish may be threatened by turbidity levels in the following ways (Sorensen et al. 1977):
(1) by acting directly on the fish swimming in water in which solids are suspended, and either killing them or reducing their growth rate, resistance to disease, etc.;
(2) by preventing the successful development of fish eggs and larvae;
(3) by modifying natural movements and migrations of fish;
(4) by reducing the abundance of food available to the fish, and

(5) by affecting the efficiency of methods of catching fish.

A Generalization

Turbidity characteristics of surface waters vary seasonally. The maximal turbidity will occur at the time of flood run off during the annual rainy season. As a general rule, turbidity in receiving waters (lagoon, estuary, lake) decreases with increasing distance from the mouth of an inflowing river; this condition obtains during every season of the year. Measurements taken in the Chesapeake Bay indicate that light intensity metered in the open Bay is the same at about ten times the depth of light measured near the mouth of the tributary Potapsco River.

With greater turbidity the photosynthetic zone of any body of water extends to lesser depths than in clearer water. Algal populations have an automatic method for curtailing size. If the abundance of algae increases greatly, the ambient water takes on a red, green, or yellow color with accompanying elevated turbidity. Photosynthesis is thus sharply curtailed and population growth declines. In certain receiving waters, breakdown of organic materials may release large amounts of tannic acid and assorted other organic acids. These substances give the waters a brown or tan hue. Examples of this phenomenon are found in polar, tropical, and temperate waters. An obvious outcome of this water staining is a reduction in the thickness of the photosynthetic layer.

BIBLIOGRAPHY

Aas, E. (1979). Light scatterance and fluorescence observations in the Barents Sea. *Inst Rept Ser Univ Oslo Inst Geofys 39*: 48 pp.

Abbott, E. A., Hiller, A. J. (1966). Some turbidimetric observations in the east Siberian Sea during July-September 1964. Naval Research Laboratory, Washington, D. C. NRL Report 6438. 5 pp.

Allen, G. P., Castaing, T., Houanneau, J. M. (1977). Mecanismes de remise en suspension et de dispersion des sediments seris dans l'estuaire de la Gironde. *Bull Soc Geol Fr 19(2)*: 167-176.

Aller, R. C., Dodge, R. E. (1974). Animal-sediment relations in a tropical lagoon Discovery Bay, Jamaica. *J Marine Res 32*: 209-232.

Amos, A. F., Gerard, R. D. (1979). Anomalous bottom water south of the Grand Banks suggests turbidity current activity. *Science 203*: 894-897.

Anderson, F. E. (1980). The variation in suspended sediment and water properties in the flood-water front traversing tidal flat. *Estuaries 3(1)*: 28-37.

Anonymous (1979). Silt curtain deployment and effectiveness. *World Dredging and Marine Construction 15(9)*: 38-42.

Apel, J. R. (1978). Past, present and future capabilities of satellites relative to the needs of ocean sciences. Bruun Memorial Lectures. UNESCO. 7-39. 19 pp.

Archer, S. (1980). A review of sensors for remote continuous automatic monitoring of water quality. Chemistry and Industry. 2 August 1980. No. 15. pp. 613-617.

Armstrong, D. E., Perry, J. J., Flatness, J. J. (1979). Availability of pollutants associated with suspended or settled river sediments which gain access to the Great Lakes. EPA-905/4-79-028. Environmental Protection Agency. Chicago, Illinois. 102 pp. National Technical Information Service. Springfield, Va. Order #PB80-189566.

Artem'ev, V. E. (1973). Carbohydrates in the suspended matter of the Pacific Ocean. *Okeanologiya. 13(5)*: 809-813.

Atlas, E., Brooks, J., Trefry, J., Sauer, T., Schwab, C., Bernard, B., Schofield, J., Giam, C. S., Meyer, E. R. (1980). Environmental aspects of ocean dumping in the western Gulf of Mexico. *J Water Pollut Control Fed 52(2)*: 329-350.

Bagnold, R. A., Barndorff-Nielsen, O. (1980). The pattern of natural size distribution. *Sedimentology 27*: 199-207.

Barretto, H. T., Summerhayes, C. P. (1975). Oceanography and suspended matter off northeastern Brazil. *J Sediment Petrol 45(4)*: 822-833.

Bartlett, G. A. (1977). Depositional environments and sediments in the coastal zone of Prince Edward Island, Canada. *Maritime Sediments 13(2)*: 41-66.

Bekasova, O. D., Kopelevich, O. V., Sud'bin, A. I. (1979). Determination of the optical properties of seawater and of the content of chlorophyll and suspended matter in the upper layer of the ocean from the brightness spectra of the upward radiation. *Oceanology 19(2)*: 145-148.

Beall, Paula T. (1981). The water of life. *The Sciences 21(1)*: 6-9, 29.

Boggs, S., Wha-Ching, W., Lewis, F. S., Chen, J-C (1979). Sediment properties and water characteristics of the Taiwan Shelf and Slope. *ACTA Oceanographica Taiwanica*. Science Reports of the National Taiwan University, No. 10. pp. 10-49.

Bohlen, W. F. et al. (1979). Suspended materials distributions in the wake of estuarine channel dredging operations. *Estuarine and Coastal Marine Science 9 (6)*: 699-712.

Bonham-Carter, G. F., Sutherland, A. J. (1967). Diffusion and settling of sediments at river mouths: a computer a simulation model. *Transactions-Gulf Coast Association of Geological Societies. 17*: 326-338.

Bruun Memorial Lectures (1978). Intergovernmental Oceanographic Technical Series. 19. UNESCO. Paris. 64 pp.

Burnes, R. E. (1980). Assessment of environmental effects of deep ocean mining of manganese nodules. *Helgoländer Meeresuntersuchungen 33*: 433-442.

Carpenter, D. J., Carpenter, S. M. (1979). A comparison of optical and biochemical classifications of ocean waters. *Deep-Sea Research 26A*: 763-773.

Castanares, A. A., Phleger, F. B., Editors, (1969). Coastal lagoons, a symposium. Universidad Nacional Autónoma de México. Ciudad Universitaria, México 20-D. F. 686 pp.

Cerofolini, G. F., Ceroffolini, M. (1980). Water structure in biosystems. *Speculations in Science and Technology 3(2)*: 149-156.

Chan, Kwan-Ming, Chen, Ju-Chin (1977). Composition of the particulate matter in the coastal waters of Taiwan. *Acta Oceanographica Taiwanica*. Science Reports of the National Taiwan University. No. 7. pp. 64-70.

Chanut, J. P., Poulet, S. A. (1979). Distribution of particle size spectra in suspension in Saguenay Fjord. *Can J Earth Sci 16(2)*: 240-249.

Chase, R. R. P. (1979). Settling behavior of natural aquatic particulates. *Limnology and Oceanography 24(3)*: 417-426.

Collins, M. B., Banner, F. T. (1979). Secchi disc depths, suspensions and circulation, northeastern Mediterranean Sea. *Marine Geology 31(1-2)*: M39-M46, April.

Council on Environmental Quality (1980). Environmental quality, 11th annual report. U.S. Gov. Printing Office, Washington, D. C. 497 pp.

Craig, H. L., Lee, T., Michel, H., Hess, S., Munier, R., Perlmutter, M. (1978). Source book of oceanographic properties affecting biofouling and corrosion of OTEC plants at selected sites. Miami University, Florida, U.S.A., Document No. PNL NTIS. Springfield, Virginia. 392 pp.

Davies, B. R. (1979). Stream regulation in Africa: A review. pp. 113-142. *In The Ecology of Regulated Streams.* Editors, J. V. Ward and J. A. Stanford. Plenum, New York. 398 pp.

Davis, A. (1973). A turbidity survey of Narragansett Bay. *Ocean Engineering 2(4)*: 169-178.

Décamps, H., J. Capblancq, H. Casanova, and J. N. Tourenq. (1979). Hydrobiology of some regulated rivers in the southwest of France. pp. 273-288. *In The Ecology of Regulated Streams* Editors, J. V. Ward and J. A. Stanford. Plenum, New York. 398 pp.

DiToro, D. M. (1978). Optics of turbid estuarine waters: approximations and applications. *Water Research 12*: 1059-1068.

Drummeter, L. F., Mertens, L. E., Editors, (1975). *Ocean Optics.* Society of Photo-optical Instrumentation Engineers. 338 Tejon Pl., Palos Verdes Estates. California. SPIE vol. 64. 160 pp.

Duinker, J. C., Nolting, R. F. (1978). Mixing, removal and mobilization of trace metals in the Rhine Estuary. *Netherlands Journal of Sea Research 12(2)*: 205-223.

Duinker, J. S., Nolting, R. F., Van Der Sloot, H. A. (1979). The determination of suspended materials in coastal waters by different sampling and processing techniques. *Netherlands Journal of Sea Research 13(2)*: 282-297.

Eichner, D., Hach, C. C. (1971). The absolute turbidity of pure water. Technical Information Series. Booklet No. 2. Hach Chemical Company. Ames, Iowa. 10 pp.

Eisma, D. (1980). Supply and deposition of suspended matter in the North Sea. *In*: Nil, S. D., Schuettenhelm, R. G. E., VanWeering, T. C. E. (eds.) *Holocene Marine Sedimentation in the North Sea Basin.* I. A. S. Special Publication V.

Eisma, D. (1981). Suspended matter as a carrier for pollutants in estuaries and the sea. *In*: Geyer, R. A. (ed.) *Marine Environmental Pollution, 2. Mining and Dumping.* Elsevier Publication Company, Amsterdam.

Eisma, D., Gieskes, W. W. C. (1977). Particle size spectra of non-living suspended matter in the southern North Sea. Interne Verslagen Nederlands Instituut voor Onderzoek der Zee. Texel. 1977-7. 22 pp.

Eisma, D., Kalf, J. (1978). Suspended matter between Norway and Shetland and in the Sognefjord. Interne Verslagen Nederlands Instituut voor Onderzoek der Zee. Texel. 1978-13. 28 pp.

Eisma, D., Kalf, J. (1979). Distribution and particle size of suspended matter in the Southern Bight of the North Sea and the Eastern Channel. *Netherlands Journal of Sea Research 13(2)*: 298-324.

Eisma, D., Kalf, J., Van der Gaast, S. J. (1978). Suspended matter in the Zaire Estuary and the adjacent Atlantic Ocean. *Netherlands Journal of Sea Research 12(3/4)*: 382-406.

Eisma, D., Van der Gaast, S. J., Martin, J. M., Thomas, A. J. (1978). Suspended matter and bottom deposits of the Orinoco Delta: turbidity, mineralogy and elementary composition. *Netherlands Journal of Sea Research 12(2)*: 224-251.

Eisenberg, D., Kauzmann, W. (1969). *The Structure and Properties of Water.* Oxford University Press. New York. 296 pp.

Ellis, K. M., Chappopadhyay, A. (1979). Multi-element determination in estuarine suspended particulate matter by instrumental neutron activation analysis. *Anal Chem 51(7)*: 942-947.

Emery, K. O., Lepple, F., Toner, Lois, Uchupi, E., Rioux, R. H., Pople, W., Hulburt, E. (1974). Suspended matter and other properties of surface waters of the northeastern Atlantic Ocean. *J Sediment Petrol 44(4)*: 1087-1110.

Environmental Protection Agency (1979). Turbidity. Method 180.1 (Nephelometric) STORET No. 00076. pp. 180.1-1 through 180.1-4. *In: Manual of Methods for Chemical Analysis of Water and Wastes*, Revised 1979, Office of Technology Transfer, Washington, D. C. 20460.

Farmsworth, E. G., M. C. Nichols, Carolyn N. Vann, Lois G. Wolfson, R. W. Bosserman, P. R. Hendrix, F. B. Golley, and J. L. Cooley (1979). *Impacts of Sediment and Nutrients on Biota in Surface Waters of the United States.* EPA-60013-79-105. Athens, Ga. 315 pp.

Faye, R. E., Carey, W. P., Stamer, J. K., Kleckner, R. L. (1980). Erosion, sediment discharge, and channel myobiology in the upper Chattahoochie River Basin, Georgia. *Geol Surv Prof Paper 1107.* GPO Wash. 85 pp.

Fenchel, T. (1980b). Suspension feeding in ciliated protozoa: feeding rates and their ecological significance. *Microbial Ecology 6*: 13-25.

Fenchel, T. (1980). Suspension feeding in ciliated protozoa: functional response and particle size selection. *Microbial Ecology 6*: 1-11.

Festa, J. F., Hansen, D. V. (1978). Turbidity maxima in partially mixed estuaries: a two-dimensional numerical model. *Estuarine and Coastal Marine Science 7(4)*: 347-359.

Fischer, J. K. (1977). Distribution of suspended particles in the equatorial region of the eastern Pacific Ocean. *Pol Archs Hydrobiol 24 (Suppl.)*: 103-108.

Fischer, J. K., Karabashev, G. S. (1977). A comparison of the size distribution of suspended particles and their optical properties. *Pol. Archs Hydrobiol 24 (Suppl.)*: 109-113.

Frank, H. S. (1970). The structure of ordinary water. *Science 169(3946)*: 635-641.

Fuguères, G., Martin, J.-M., Meybeck, M. (1978). Iron behaviour in the Zaire Estuary. *Netherlands Journal of Sea Research 12(3/4)*: 329-337.

Gardner, W. D. (1977). Fluxes, dynamics, and chemistry of particulates in the ocean. Dissertation for the Ph.D. Massachusetts Institute of Technology and Woods Hole Oceanographic Institute. 405 pp. Library, Marine Biological laboratory, Number GC 7.1 G 17.

Garfield, P. C., Packard, T. T., Codispoti, L. A. (1979). Particulate protein in the Peru upwelling system. *Deep-Sea Res, 26(6A)*: 623-639.

Gerrodette, T., Flechsig, A. O. (1979). Sediment-induced reduction in the pumping rate of the tropical sponge *Verongia lacunosa*. *Marine Biology 55*: 103-110.

GESAMP (1975). Joint group of experts on the scientific aspects of marine pollution. Report of the Seventh Session. London. 24-30 April, 1975. Inter-governmental Maritime Consultative Organization. IMCO/FAO/UNESCO/WMO/WHO/IAEA/UN. 12 pp. plus 9 annexes.

Ghovaniow, A. H., Hickman, G. D., Hogg, J. E. (1973). Laser transmission studies of East Coast waters. Contract Rpt. ONR N00014-71-C-0202. March. 41 pp. Order No. AD-759 781. National Technical Information Service, Virginia.

Gierloff-Emden, H. G. (1977). *Orbital Remote Sensing of Coastal and Offshore Environments*. A manual of interpretation, Walter de Gruyter, Berlin. 176 pp.

Gilbert, R. (1978). Observations on oceanography and sedimentation at Pannirtung Fiord, Baffin Island. *Maritime Sediments 14(1)*: 1-9.

Gregg, R. E., Bergersen, E. P. (1980). *Mysis relicta*: Effects of turbidity and turbulence on short-term survival. *Trans Am Fish Soc 109*: 207-212.

Gordon, H. R., Smith, J. M. (1972). Light-scattering profiles in the Straits of Florida. *Bull Mar Sci 22(1)*: 1-9.

Hails, J. R. (1974). A review of some current trends in nearshore research. *Earth-Science Rev 10*: 171-202.

Harris, J. E., Fabris, G. P. (1979). Concentrations of suspended matter and particulate cadmium, copper, lead and zinc in the Indian sector of the Antarctic Ocean. *Mar Chem 8(2)*: 163-180.

Harvey, H. W. (1960). *The Chemistry and Fertility of Seawaters*, Cambridge University Press. 240 pp.

Hach, C. C. (1979). Introduction to turbidity measurement. Technical Information Series-Booklet No. 1, 2nd edition. Hach Chemical Co. Loveland, Colorado. 8 pp.

Helliwell, P. R., Bossanyi, J., Editors, (1975). *Pollution Criteria for Estuaries*. John Wiley & Sons. New York. 302 pp.

Hildreth, D. I., Mallet, A. (1980). The effect of suspension density on the retention of 5 M diatoms by the *Mytilus edulis* gill. *Biological Bull 158*: 316-324.

Hinga, K. R., Sieburth, J. McN., Heath, G. R. (1979). The supply of organic material at the deep sea floor. *J Mar Res 37(3)*: 557-579.

Holmes, R. W. (1957). Solar radiation. Ch. 6. pp. 109-128. *In*: Memoir no. 67. *Treatise on Marine Ecology and Paleoecology*. Vol. 1, Ecology. Geological Society of America, New York, 1296 pp.

Holyer, R. J. (1978). Toward universal multispectral suspended sediment algorithms. *Remote Sensing of Environment 7*: 323-338.

Holyer, R. J. (1980). Multispectral techniques for remote monitoring of sediment in water: a feasibility investigation. EPA-600/4-80-019. Contract EPA-68-03-2153. Lockheed Electronics Co., Inc., Las Vegas, NV. Remote Sensing Lab. 173 pp.

Honjo, S. (1980). Material fluxes and modes of sedimentation in the mesopelagic and bathypelagic zones. *J Marine Res 38(1)*: 53-97.

Hubbell, D. W., Glenn, J. L., Stevens, H. H. (1971). Studies of sediment transport in the Columbia River Estuary. Proceedings. Technical Conference on Estuaries of the Pacific Northwest. Circular No. 42. Engineering Experiment Station. Oregon State University. Corvallis. pp. 190-226.

Hunter, K. A. (1980). Processes affecting particulate trace metals in the sea surface microlayer. *Mar Chem 9*: 49-70.

Ignatiad, L., Becacosk, T. (1970). Ecology of fouling organisms in a polluted area. *Nature (London) 225*: 293-294.

Inglis, C. C., Allen, F. H. (1957). The regime of the Thames Estuary as affected by currents and river flow. *Proc Inst Civ Engin Lond 7*: 827-868.

Iwamota, R. N., Salo, E. O., Madej, M. A., McComas, R. L., Rulifson, R. L. (1978). Sediment and water quality: a review of the literature including a suggested approach for water quality criteria with summary of workshop and conclusions and recommendations. ETA910/9-78-048. 151 pp.

Jerlov, N. G. (1968). *Optical Oceanography*. Elsevier. New York. 197 pp.

Jerlov, N. G., Nielsen, E. S. (1974). *Optical Aspects of Oceanography*. Academic. New York. 494 pp.

Jørgensen, C. B. (1975). On gill function in the mussell *Mytilus edulis L. Ophelia 13*: 187-232.

Juda, L. (1979). The regional effort to control pollution in the Mediterranean Sea. *Ocean Management 5*: 125-150.

Kamenkovich, V. M., Monin, A. S., Editors, (1978). *Oceanology. Physics of the Ocean*. 1. Hydrophysics of the ocean. Nauka, Moscow. 476 pp.

Kasymov, A. G., Likhodeyeva, N. F. (1979). Filtration capacity of some molluscs of the Caspian Sea. *Water, Air, and Soil Pollution 11*: 279-288.

Kiorboe, T., E. Frantsen, C. Jensen, and G. Sorensen (1981).Effects of suspended sediment on development and hatching of herring (*Clupea harengus*) eggs. *Estuarine, Coastal and Shelf Science 13(1)*: 107-111.

Kirk, J. T. O. (1979). Spectral distribution of photosynthetically active radiation in some south-eastern Australian waters. *Australian Journal of Marine and Freshwater Research 30(1)*: 81-91, Feb.

Kondratev, K. Y., Buznidov, A. A., Pozdnyakov, D. V. (1972). Remote sensing of pollution of water basins and phytoplankton by optical methods. Water Resources. No. 3.

Kozlyaninov, M. V. (1974). Optical characteristics in the bottom layers of the ocean. *Oceanology 14(6)*: 824-826. English Transl. Serv. Amer. Geophys. Union.

Kozlyaninov, M. V. (1979). On the luminance coefficient of sea water. *Oceanology 19(2)*: 221-227.

Kozlyaninov, M. V. (1979). On the brightness coefficient of the sea. *Oceanology 19(2)*: 136-140.

Kullenberg, G. (1975). Dispersion of fine-grained material and other physical aspects. Appendix II to Annex IX. *In*: Report of Seventh Session. Joint Group of Experts on the Scientific Aspects of Marine Pollution, No. 1, Intergovernmental Maritime Consultative Organization, London, GESAMP, VII/9.

Kuo, A., Nichols, M., Lewis, J. (1978). Modeling sediment movement in the turbidity maximum of an estuary. Virginia Polytechnic Institute, Blacksburg. *Water Resources Research Center Bull* No. 11. 82 pp., June.

Kuroki, T., Editor, (1975). Preliminary report of the Hakuho Maru cruise K. H-7A-2 (N. N. Pacific Cruise). April 30-June 26, 1974. Ocean Research Institute. University of Tokyo, Japan. 72 pp.

Lam, R. K., Frost, B. W. (1976). Model of copepod filtering response to changes in size and concentration of food. *Limnol Oceanog 21*: 490-500.

Lehman, J. T. (1976). The filter-feeder as an optimal forager, and the predicted shapes of feeding curves. *Limnol Oceanog 21*: 501-516.

Loya, Y. (1976). Effects of water turbidity and sedimentation on the community structure of Puerto Rican corals. *Bull Marine Science 26*: 450-466.

Luening, K., Dring, M. J. (1979). Continuous underwater light measurement near Helgoland (North Sea) and its significance for characteristic light limits in the sublittoral region. *Helgolaender wiss Meeresunters 32*: 403-424.

Martin, J.-M., Meybeck, M., Pusset, M. (1978). Uranium behaviour in the Zaire Estuary. *Netherlands Journal of Sea Research 12(3/4)*: 338-344.

Martin, J.-M., Thomas, A. J., van Grieken, R. (1978). Trace element composition of Zaire suspended sediments. *Netherlands Journal of Sea Research 12(3/4)*: 414-420.

McCave, I. N. (1978). Suspended material over the central Oregon continental shelf in May 1974: I, concentrations of organic and inorganic components. *Journal of Sedimentary Petrology 49(4)*: 1181-1194.

Meade, R. H. (1972). Sources and sinks of suspended matter on continental shelves. *In*: Swift, Duane, Pilkey (ed.) *Shelf Sediment Transport*, Ch. 11, Dowden, Hutchinson & Ross, Stroudsburg, Pa., pp. 249-262.

Meade, R. H., Sachs, P. L., Manheim, F. T., Hathaway, J. C., Spencer, D. W. (1975). Sources of suspended matter in waters of the Middle Atlantic Bight. *J Sedimentary Petrology 45(1)*: 171-188.

Miller, J. M. (1974). Nearshore distribution of Hawaiian marine fish larvae: effects of water quality, turbidity and currents. *In*: Blaxter, J. H. S. (ed.), *The Early Life History of Fish*, Springer-Verlag, New York, 765 pp.

Miller, J. R., Jain, S. C., O'Neill, N. T., McNeil, W. R., Thomson, K. P. B. (1977). Interpretation of airborne spectral reflectance measurements over Georgian Bay. *Remote Sensing of Environment 6*: 183-200.

Milliman, J. D., Summerhayes, C. P., Barretto, N. T. (1975). Oceanography and suspended matter off the Amazon River, February-March, 1973. *J Sed Petrol 45*: 189-206.

Minagawa, M., Tsunogai, S. (1980). *Earth and Planetary Science Letters 47(1)*: 51-64.

Møhlenberg, F., Riisgård, H. U. (1978). Efficiency of particle retention in 13 species of suspension feeding bivalves. *Ophelia 17(2)*: 239-246.

Monniot, C. (1979). Adaptations of benthic filtering animals to the scarcity of suspended particles in deep water. Ambio Special Report. No. 6, pp. 73-74. Royal Swedish Academy of Sciences, Stockholm, 104 pp.

Monniot, C., Monniot, F. (1977). Tuniciers benthiques profonds du Nord-Est Atlantique. Resultats des campagnes Biogas. *Bull Mus natn Hist nat Paris 3eme Sér. 466. Zool. 323*: 695-719.

Moore, P. G. (1977). Inorganic particulate suspensions in the sea and their effects on marine animals. *Oceanogr Mar Biol Ann Rev 15*: 225-363.

Morel, Francois, Swallow, D. C. (1980). Adsorption of trace metals by hydrous ferric oxide in seawater. Final rept. EPA-600/3-80-011, Jan 80, 63 pp.

Morton, J. W. (1977). Ecological effect of dredging and dredge spoil disposal: a literature review. *U.S. Fish and Wildl Serv Techn Paper 94*: 33 pp.

Muncy, R. J., Atchison, G. J., Bulkley, R. V., Menzel, B. W., Perry, L. G., Summerfelt, R. C. (1979). Effects of suspended solids and sediment on reproduction and early life of warm water fishes: a review. Corvallis Environmental Research Laboratory, Office of Research and Development, U.S. Environmental Protection Agency, EPA-600/3-79-042, Corvallis, Oregon, 110 pp.

Nakai, O. (1978). Turbidity generated by dredging projects. *In*: Peterson, S. A., Randolph, K. K. (ed.), *Management of Bottom Sediments Containing Toxic Substances*: proceedings, U.S. Environmental Protection Agency, Off. R. & D., Ecol. Res. Ser. EPA-600/3-78-084, pp. 31-47, (NTIS).

Nakajima, K. (1969). Suspended particulate matter in the waters on both sides of the Aleutian Ridge. *Oceanographical Soc Japan (25(5))*: 239-248.

Nakajima, K. (1973). Suspended particulate matter in the western north Pacific Ocean. *Mem Fac Fish, Hokkaido Univ 20 (1-2)*: 1-106.

Nakajima, K., Nishizawa, S. (1972). Exponential decrease in particulate carbon concentration in a limited depth interval in the surface layer of the Bering Sea. *In*: Takenouti, A. Y. and others (ed.), *Biological Oceanography of the Northern North Pacific Ocean*, Idemitsu Shoten, Tokyo, Japan. pp. 495-505.

Nelepo, B. A. (1978). Remote sensing of the ocean in the USSR. Bruun Memorial Lectures, UNESCO, 19 pp., 41-50.

Nelson, L. A. (1979). Minor elements in the sediments of the Thames Estuary. *Estuarine and Coastal Marine Science 9(5)*: 623-629.

Neumann, G., Pierson, W. J. (1966). *Principles of Physical Oceanography*. Prentice-Hall, Englewood Cliffs, 545 pp.

Nihoul, J. C. J. (ed.) (1978). *Hydrodynamics of Estuaries and Fjords*. Elsevier Oceanography Series 23, Elsevier Scientific Publishing Co., Amsterdam, 546 pp.

Nishizawa, S., Nakajima, K. (1971). Concentration of particulate organic material in the sea surface layer. *Bull Plankton Soc Japan 18(2)*: 12-19.

Optical Society of America (1966). Science of color. Committee on Colorimetry, 6th ed., Washington, D.C.

Palmer, R. E. (1980). Behavioral and rhythmic aspects of filtration in the bay scallop, *Argopecten irradians concentricus* (Say), and the oyster, *Crassostrea virginica* (Gmelin). *Journal of Experimental Marine Biology and Ecology 45*: 273-295.

Pantin, H. M. (1979). Interaction between velocity and effective density in turbidity flow: phase-plane analysis, with criteria for autosuspension. *Marine Geology 31(1-2)*: 59-99, April.

Pelevin, V. N., Gurfink, A. M., Gol'din, Yu A. (1979). The effect of the form of scattering indicatrix on the non-stationary light field in the sea. *Oceanology 19(2)*: 228-232.

Pelevin, V. N., Rutkovskaya, V. A. (1978). Attenuation of photosynthetically active solar radiation in the waters of the Pacific Ocean. *Oceanol Acad Sci USSR 18(4)*: 408-412.

Penhoat, Y. du, Salomon, J. C. (1979). Numerical simulation of estuarine turbidity maximum. Application to the Gironde estuary. *Oceanol Acta 2(3)*: 253-260.

Pierce, J. W., Siegel, F. R. (1979). Particulate material suspended in estuarine and oceanic waters. *Scanning Electron Microsc 1979(1)*: 555-562.

Plank, W. S., Zaneveld, J. R. V., Pak, H. (1973). Distribution of suspended matter in the Panama Basin. *Geophysical Research 78(30)*: 7113-7121.

Poulet, S. A. (1973). Grazing of *Pseudocalanus minutus* on naturally occurring particulate matter. *Limnology and Oceanography 18(4)*: 564-573.

Preisendorfer, R. W. (1976). *Hydrologic Optics*, Vol. I, Introduction. Vol. II, Foundations. Vol. III, Solutions. Vol. IV, Imbeddings. Vol. V, Properties. U.S. Department of Commerce, National Oceanic and Atmospheric Administration Environmental Research Laboratories, Pacific Marine Environmental Laboratory, Honolulu, Hawaii, pp. 218, 400, 245, 207, 296 respectively.

Probst, B. (1976). A model to simulate the cycling of pelagic material in a marine shallow water ecosystem of the Western Baltic. Kieler Meeresforschungen Nr 3 pp.

Proni, J. R., Rona, D. C., Lauter, C. A., Sellers, R. L. (1975). Acoustic observations of suspended particulate matter in the ocean. *Nature 254(5499)*: 413-415.

Pustel'nikov, O. S., Urbanovich, I. M. (1974). Distribution of carbohydrates and lipids in suspended matter in the north Atlantic Ocean. *Okeanologiya 14(4)*:649-654.

Raisbeck, G. N., Yiou, F. U., Fruneau, M., Loiseaux, J. M., Lieuvin, N. (1979). *Earth and Planetary Science Letters 43(2)*: 237-240.

Rao, V. C., Rao, T. S. S. (1975). Distribution of particulate organic matter in the Bay of Bengal. *J Mar Biol Assn, India 17(1)*: 40-55.

Reimnitz, B., Barnes, P. W. (1973). Studies of the inner self and coastal sedimentation environment of the Beaufort Sea from ERTS-1. U.S. NASA, Contract Report NASA-CR-132240, Order Number E73-10649, 6 pp.

Rhoads, D. C. (1974). Organism-sediment relations on the muddy sea floor. *Marine Biol Ann Rev 12*: 263-300.

Rhoads, D. C., Young, D. K. (1971). Animal-sediment relations in Cape Cod, Massachusetts. Part II. Reworking by *Molpadia oolitica* (Holothuroidea); *Marine Biology 11*: 255-261.

Riisgård, H. U., Randlov, A., Kristensen, P. S. (1980). Rates of water processing, oxygen consumption and efficiency of particle retention in veligers and young post-metamorphic *Mytilus edulis*. *Ophelia 19(1)*: 37-47.

Roessler, W. G., Brewer, C. R. (1967). Permanent turbidity standards. *Applied Microbiology 15(5)*: 1114-1121.

Rosenthal, H., Alderdice, D. F. (1976). Sublethal effects of environmental stressor, natural and pollutional, on marine fish eggs and larvae. *J Fish Res Board Can 33(9)*: 2047-2065.

Ross, D. A. (1977). *Introduction to Oceanography*. Second edition. Prentice-Hall, Englewood Cliffs, 438 pp.

Sagdeev, R. Z. (1977). Investigating the earth from space. *Bull Acad Sciences of the USSR*, No. 3.

Saino, T., Hattori, A. (1980). ^{15}N natural abundance in oceanic suspended particulate matter. *Nature 283(5749)*: 752-754.

Self, R. F. L., Jumars, P. A. (1978). New resources axes for deposit feeders. *J Mar Res 36(4)*: 627-641.

Servizi, J. A. (1971). A study of sediments from Bellingham Harbor as related to marine disposal. Proceedings Technical Conference on Estuaries of the Pacific Northwest, Circular No. 42, Engineering Experiment Station, Oregon State University, Corvallis, pp. 227-248.

Sheldon, R. W., Prakash, A., Sutcliff, W. H. Jr. (1972). The size distribution of particles in the ocean. *Limnology and Oceanography 17*: 327-340.

Shepherd, I. E. (1978). A silt-salinity-depth profiling instrument. *Dock and Harbour Authority 59(690)*: 9-10, May.

Sherk, J. A., Cronin, L. E. (1970). The effects of suspended and deposited sediments on estuarine organisms. An annotated bibliography. Chesapeake Biological Laboratory, Solomons, Maryland, Natural Resources Institute, University of Maryland, Reference No. 70-19, 62 pp.

Shideler, G. L. (1979). Regional surface turbidity and hydrographic variability of the south Texas continental shelf. *Sedimentary Petrology 49(4)*: 1195-1208.

Shiroto, A. (1979). Studies on non-living particulate suspensions - II: Turbidity as an index of water masses in the estuary. *Nikon Suisan-Gakkai Shi 45(9)*: 1129-1135.

Sholkovitz, E. R. (1979). Chemical and physical properties controlling the chemical composition of suspended material in the River Tay estuary. *Estuar Coast Mar Sci 8(6)*: 523-545.

Sholkovitz, E. R., Price, N. B. (1980). The major-element chemistry of suspended matter in the Amazon Estuary. *Geochemica et Cosmochimica Acta 44*: 163-171.

Sholkovitz, E. R., van Grieken, R., Eisma, D. (1978). The major-element composition of suspended matter in the Zaire River and estuary. *Netherlands Journal of Sea Research 12(3/4)*: 407-413.

Siezen, R. J., Mague, T. H. (1978). Amino acids in suspended particulate matter from oceanic and coastal waters of the Pacific. *Mar Chem 6(3)*: 215-232.

Smetacek, R., Hendrikson, P. (1979). Composition of particulate organic matter in Kiel Bight in relation to phytoplankton succession. *Oceanol Acta 2(3)*: 287-298.

Smith, R. C., Baker, Karen S. (1978b). The bio-optical state of ocean waters and remote sensing. *Limnology and Oceanography 23(2)*: 247-259.

Smith, R. C., Baker, Karen S. (1978). Optical classification of natural waters. *Limnology and Oceanography 23(2)*: 260-267.

Smith, J. D., Milne, T. J. (1979). Determination of iron in suspended matter and sediments of the Yarra River Estuary, and the distribution of copper, lead, zinc and manganese in the sediments. *Aust J Mar Freshwater Res 30*: 731-739.

Socci, A., Tanner, W. F. (1980). Little known but important papers on grain-size analysis. *Sedimentology 27*: 231-232.

Sorenson, D. L., McCarthy, M. M., Middlebrooks, E. J., Porcella, D. B. (1977). Suspended and dissolved solids effects on freshwater biota: A review. Environmental Protection Agency, Office of Research and Development, Res. Rept. ETA-600/3-77-042, 64 pp.

Sugihara, S., Inoue, N. (1976). Measurements of spectral energy distribution in the sea. *Scientific Papers of the Institute of Physical and Chemical Research 70*: 8-15.

Sugihara, S., Kawana, K. (1975). Polarization of the scattered light by turbid sea water and the size distribution of suspended particles. *Scientific Papers of the Institute of Physical and Chemical Research 69*: 6-15.

Sugihara, S. (1977). Influence of scattered light on underwater light field. *Scientific Papers of the Institute of Physical and Chemical Research 71*: 1-2.

Sugihara, S., Tsuda, R. (1979a). Size distribution of suspended particles in the surface water of the central North Pacific Ocean. *Scient Pap Inst Phys Chem Res, Tokyo 73(1)*: 1-8.

Sugihara, S., Tsuda, R. (1979b). Light scattering and size distribution of particles in the surface waters of the North Pacific Ocean. *J Oceanogr Soc, Japan 35(2)*: 82-90.

Swenson, W. A. (1978). Influence of turbidity on fish abundance in western Lake Superior. EPA-600/3-78-067, July 1978, Grant Member R802455 Environmental Research Laboratory — Duluth, Office of Research and Development, U.S. Environmental Protection Agency, Duluth, MN., 55804, 92 pp.

Taghon, G. L., Self, R. F. L., Jumars, P. A. (1978). Predicting particle selection by deposit feeders: a model and its implications. *Limnol Oceanog 23(4)*: 752-759.

Taguchi, S., Nakajima, K. (1971). Plankton and seston in the sea surface of Three Inlets of Japan. *Bull Plankton Soc Japan 18(2)*: 1-36.

Teal, J., Teal, Mildred (1975). *The Sargasso Sea*. Atlantic Monthly Book, Little, Brown and Co., Boston, 216 pp.

Tevesz, M. J. S., Soster, F. M., McCall, P. L. (1980). The effects of size-selective feeding by oligochaetes on the physical properties of river sediments. *J Sedimentary Petrology 50(2)*: 561-568.

Theisen, B. F. (1977). Feeding rate of *Mytilus edulis* L. (Bivalvia) from different parts of Danish waters in water of different turbidity. *Ophelia 16(2)*: 221-232.

Thiel, R. (1975). The size structure of the deep sea benthos. *Intern Rev Ges Hydrobiol 60*: 575-606.

Thompson, M. J., Gilliland, L. E., Rosenfield, L. K. (1979). Light scattering and extinction in a highly turbid coastal inlet. *Estuaries 2(3)*: 164-171.

Timofeeva, V. A. (1961). On the problem of the polarization of light in turbid media. *Izv Akad Nauk SSSR Ser Geofiz 5*: 766-774.

Turekian, K. K. (1977). The fate of metals in the oceans. *Geochimica et Cosmochimica Acta 41*: 1139-1144.

Tyler, J. E., Smith, R. C. (1970). *Measurements of Spectral Irradiance Underwater*. Gordon and Breach, New York, 103 pp.

Vandenberghe, J. A. (1978). Submarine radiometry and physical environment monitoring. In: Oceans '78: The ocean challenge, Proc. 4th Ann. MTS/IEEE Conference, pp. 571-574. Marine Technology Society, Washington, D.C., 766 pp.

Van Hylckama, T. E. A. (1979). Water, something peculiar. *Hydrological Sciences 24(4)*: 499-507.

Vishnudatta, M. N., Murty, A. V. S. (1978). Transparency meter to study the underwater light penetration in the sea. *Indian J Marine Sciences 7(3)*: 205.

Waldichuk, M. (1977). Global marine pollution: an overview. Intergovernmental Oceanographic Commission, Technical Series 18, UNESCO, Paris, France, 96 pp.

Ward, J. V. (1974). A temperature-stressed stream ecosystem below a hypolimnial release mountain reservoir. *Arch Hydrobiol 74*: 274-275.

Ward, J. V. and J. A. Stanford. (1979). Editors. *The Ecology of Regulated Streams*. Plenum, New York. 398 pp.

Webster, J. R., E. F. Benfield, and J. Cairns, Jr. (1979). Model predictions of effects of impoundment on particulate organic matter transport in a river system. pp. 339-364. *In The Ecology of Regulated Streams*. Editors, J. V. Ward and J. A. Stanford. Plenum, New York. 398 pp.

Wellershaus, S. (1981). Turbidity maximum and mud shoaling in the Weser Estuary *Arch Hydrobiol 92(2)*: 161-198.

Wiebe, P. H., Madin, L. P., Haury, L. R., Harbison, G. R., Philbin, L. M. (1979). Diel vertical migration by *Salpa aspersa* and its potential for large-scale particulate organic matter transport to the deep-sea. *Mar Biol 53*: 249-255.

Wilber, C. G. (1964). Animals in aquatic environments. In: Dill, D. B. Adolph, E. F., Wilber, C. G. (ed.) Adaptation to the environment. *Handbook of Physiology*, pp. 661-668, Williams & Wilkins, Baltimore, 1056 pp.

Wilber, C. G. (1969). *The Biological Aspects of Water Pollution*. Charles C Thomas, Publisher, Springfield. 296 pp.

Wilber, C. G. (1971a). Turbidity. General introduction. In: Kinne, O. (ed.) *Marine Ecology*, Vol. I, part 2, ch. 6.0. Wiley-Interscience, New York, pp. 1157-1165.

Wilber, C. G. (1971b). Turbidity. Animals. In: Kinne, O. (ed.) *Marine Ecology*, Vol. I, part 2, ch. 6.3. Wiley-Interscience, New York, pp. 1181-1194.

Williams, J. (1970). *Optical Properties of the Sea*. United States Naval Institute, Annapolis, MD., 123 pp.

Winter, J. E. (1969). Uber den Einfluss der Nahrungskonzentration und anderer Faktoren auf Filtrierleistung und Nahrungsausnutzung der Muscheln *Arctica* islandica und *Modiolus modiolus*. *Marine Biology 4*: 87-135.

Yangawa, S., Kashiwa, T., Inoue, K. (1978). Studies on the formation of fishing ground around Yamato Bank. Part I. Oceanographic structure. Tokyo Univ. Fish. 5-7, Konan 4, Minato, Tokyo 108, *J P N Mar Tokyo 16(1)*: 23-35.

Zeitzschel, B. (1970). The quantity, composition and distribution of suspended particulate matter in the Gulf of California. *Marine Biology 7*: 305-318.

INDEX

A
Absorbance, 103
Absorption, 77
  of metals, 71-72
Africa, 97-98
African coast, 97
Amazon Estuary, 87
Amino acids, 71
Aquatic invertebrates, 41
  organisms, 27
Asexual reproduction, 54
Attenuation, 7, 11, 62
  coefficient, 8, 15, 103
  length, 8
Australia, 103

B
Baffin Island, 89-90
Bay of Bengal, 91
Beaufort Sea, 90-91
Bering Sea, 92-93
Bertalanffy's equation, 54
Biological effects, 25
Brightness coefficient, 6
Bristol Bay, 100

C
Canada, 101
Characteristic absorption length, 15
Clay particles, 34
Color photography, 58, 60-61
Columbia River, 81
Continuous monitoring, 57
Coral communities, 56
Corals, 20

D
Dams, 16-17
Data interpretation, 58-59
Deep sea mining, 39-40
Definitions, 3
Deposit feeders, 35
Deposited material, 54
Distribution of larvae, 54-55
Dredging, 21-22, 40, 47-50
Dynamic aspects of turbidity, 115
Dynamics of suspended particles, 67-72

E
Ecological effects of turbidity, 18
Effects of turbidity, 23
Embryonic development, 49
Esthetics of turbidity, 112-113
Estuaries, 22, 73-88
Estuarine water quality, 74
Estuary, definition, 73
Excessive turbidity, 48
Extinction coefficient, 77

F
Fate of metals in estuaries, 87-88
Feeding, 28
  efficiency, 29
  rates, 30-31, 33-36
  selection, 30
Filter feeder feeding, 31-33
Filtration speed, 33
Fish development, 39
Florida, 77-78
Flux data, 19
Formazin turbidity units (FTU), 20, 62-63
Fort Pierce Inlet, 77-78

Fresh water data, 115-116
  geographic studies, 107-108
  invertebrates, 27, 41
  quality, 113-114
  shrimp, 28
  studies, 47
Fresh water suspended solids, 20
  turbidity, 66

G

Geographic studies, 89-108
Gironde River, 76
Government recommended turbidity methods, 65
Great Lakes, turbidity, 45-47
Gulf of California, 106

H

Hach turbidimeter, 64
Helgoland, 107
Hellige turbidimeter, 102

I

Illumination, 58
Instrumentation, 64-65
Irradiance, 8, 77

J

Jackson turbidity units (JTU), 64

L

Lagoons, 86-87
Lake Superior, 53
Larvae, 54-55
Larval stages, 47
Lethal levels, 46
Light energy and ocean depth, 13
Light measurements, 9
  penetration, water, 59-60
  loss of intensity, 10

M

Man made turbidity, 111-112
Mathematical modelling, 76-77
Maxima and minima, 78
Mead Lake, 62
Mediterranean pollution, 100-101
Metal adsorption, 71-72
Metals in estuaries, 87-88
Methodology, 57

Methods, 63-64
Mid-Atlantic Bight, 91-92
Migrations and particulate matter, 19
Monitoring, 57, 62-63, 80
Mussels, 36

N

Naragansett Bay, 76
National estuary water quality, 74
Nepheloid layer, 68-69
Nephelometric turbidity units (NTU), 64
Nephelometry, 64
Non-salmonid fish, 44
North Pacific Ocean, 93-97
North Sea, 106

O

Ocean disposal, 109-111
  mining, 38
Oceanic particles, 68
  suspended material, 99-100
Offshore transport, 98-99
Optical characteristics, ocean, 7, 14, 58
Optical studies, ocean, 4

P

Pacific Ocean, 93-97
Panama, 102
Particle density, 16
  retention, 35
  settling, 13, 14
  size, 9, 80
Particles, oceanic, 18
Particulate matter, 3, 38
Particulate matter and migrations, 19
Particulate organic matter, 70
Particulate scavenging, 69
Photographic imagery, 60-62
Photosynthesis, 103
Physical considerations, 3
Physiological effects, 26-27
Polarization, 66
Pollution, 36, 37, 56, 80-81, 100-101
Properties of water, 5
Protozoa, 32
Pumping rate, 29

R

Radio-tracer studies, 76
Rayleigh ratio, 5, 6

Reflectance, 15, 76
Reflection of sunlight, 12
Refractive index of water, 5
Remote sensing, 57-62
Reproduction, 46
Reproductive movements, 48
Residence times, 37
Resuspension, 67-68
Rhine River, 87
Runoff, 50-51

S
Salmonid fish, 51-54
Sand and gravel, 40
Sargasso Sea, 97
Scatterance, 106
Scattering, 77
    of light, 5, 6
Secchi disk, 90, 104
Sediment classification, 4
Seston, 25
Shoreline, 39
Siberia, 102-103
Spanish Coast, 103-104
Specific effects, 27
Stoke's Law, 13
Submarine radiometry, 63
Suspended amino acids, 71
Suspended material in estuaries, 79-80
Suspended matter, 17
Suspended particulate matter, 19, 27
Suspended particulates, 25
Suspended sediments, 3, 43
Suspended solids, 14, 24, 28, 47

T
Taiwan, 107

Tay Estuary, 80
Trace elements, 16
Transparency, 58
Turbidimeter, 64
Turbidity, decreased, 52-56
    in streams, 17
    methods, 65
    research, 114-115
    standards, 57
    units, 80
    values, 57
Turbidity and fish, 42-45, 50-53
Turbidity characteristics, 116
Turbidity current, 4

U
Uranium, 85-86

V
Vertical fluxes, 18
Vision in fish, 47

W
Water characteristics, 73
Water disposal, 109
Water quality, 22, 47, 48
Weser, 78-79
Whatcom Waterway, 78-79, 81-82
Worms, 35

Y
Yarra, 73-75
Yeast cells, 34

Z
Zaire Estuary, 83-86
Zaire River, 82-84